Emerging Research in Materials for Environment, and Civil Infrastructure
GeoME 5.5

International Scientific Conference on Geosciences and Environmental Management (GeoME 5.5), held at the Higher School of Technology in Salé, Morocco, from 16 to 18 October 2025

Editor
Khadija BABA[1], Abderrahman NOUNAH[1], Latifa OUADIF[2], Lahcen BAHI[2], Younes EL RHAFFARI[1]

[1]Higher School of Technology of Sale, Morocco
[2] Mohammadia Engineering School, Mohammed V University of Rabat

Peer review statement

All papers published in this volume of "Materials Research Proceedings" have been peer reviewed. The process of peer review was initiated and overseen by the above proceedings editors. All reviews were conducted by expert referees in accordance to Materials Research Forum LLC high standards.

Published under License by **Materials Research Forum LLC**
Millersville, PA 17551, USA

Published as part of the proceedings series
Materials Research Proceedings
Volume 58 (2026)

ISSN 2474-3941 (Print)
ISSN 2474-395X (Online)

ISBN 978-1-64490-392-6 (Print)
ISBN 978-1-64490-393-3 (eBook)

This book contains information obtained from authentic and highly regarded sources.
Reasonable efforts have been made to publish reliable data and information, but the
author and publisher cannot assume responsibility for the validity of all materials or the
consequences of their use. The authors and publishers have attempted to trace the
copyright holders of all material reproduced in this publication and apologize to
copyright holders if permission to publish in this form has not been obtained. If any
copyright material has not been acknowledged please write and let us know so we may
rectify in any future reprint.

Distributed worldwide by

Materials Research Forum LLC
105 Springdale Lane
Millersville, PA 17551
USA
https://mrforum.com

Manufactured in the United State of America
10 9 8 7 6 5 4 3 2 1

Table of Contents

Preface

Committees

Comparative seismic vulnerability of shear walls and moment frames in RC buildings
Sana Afnzar, Mohamed Ahatri, Mina Derife, Abderrahman Atmani, El Hassan Ait Laasri 1

Concrete strength prediction: Exploring an ensemble machine learning approach
Younes Alouan, Seif-Eddine Cherif, Badreddine Kchakech, Youssef Cherradi,
Azzouz Kchikach 9

Numerical study and buckling behavior of web-posts in stainless-steel cellular beams
Imane Bachguar, Ouadia Mouhat, Rabee Shamass, Fatima El Mennaouy 17

Optimization of concrete durability using artificial intelligence: Modeling and prediction of eco-concrete performance with fly ash
Kaoutar BAZZAR, Salma CHRIT, Adil HAFIDI ALAOUI 24

Comparative study of the mechanical and hydro-thermal behavior of earth blocs stabilized by cement or plaster and pozzolan
Imane Daya, Jamila Elbrahmi, Nouzha Lamdaour, Toufik Cherradi 32

Seismic evaluation of compressed earth blocks using acoustic waves: Case of Chichaoua Taroudant
Abderrahmane Jouhar, Mohammed Cherraj, Mokhfi Takarli, Fatima Allou, Driss El Hachmi ... 40

Comparative analysis of the stability performance of solid and perforated brick masonry structures
Chaimae Khanfri, Ouadia Mouhat, Younes El Rhaffari, Fatima El Mennaouy 47

Thermal performance of plaster composites reinforced with date palm fibers: experimental and numerical analysis
Youssef Khrissi, Mohamed Char, Amine Tilioua, Mohamed Azrour 54

3D concrete printing technology: Progress and prospects for sustainable construction
Nada Oulad Moussa, Mohamed El Haim, Loubaba Rida 62

Experimental and simulation study of the mechanical properties of eco composite materials reinforced with multiple fibers
Asmae OUROUI, Youssef BIBRIDNE, Najma LAAROUSSI, Mohammed AIT EL FQIH,
Mohammed ELWAZNA 70

Comparative study of the properties of mortar incorporating sugarcane bagasse ash and natural pozzolan
Fatima BOUKABOUS, Jalal IKEN, Omar DADAH, Khalil NACIRI, Issam AALIL,
Ali CHAABA 78

The use of timber to enhance seismic performance in vernacular architecture of Morocco: The case of the Haouz Valley
Assia LAMZAH, Ismail MAKHAD 86

Improving the non-destructive evaluation of in-situ concrete strength: The role of core location selection
Bouchra Kouddane, Youssef Jamil 92

Influence of natural additives on the heat transfer properties of clay–plaster building materials
Soufian Omari, Najma Laaroussi, Aziz Ettahir 99

New challenges in the field of heritage conservation and restoration
Daniela Pittaluga 108

The urban void of the Mellah in Sefrou's Medina, reflections through integrated digital methodologies
Giovanni Pancani, Alberto Pettineo, Giovanni Minutoli, Houssem Dine Kouidhi 116

Fast-survey techniques for digital mapping and understanding the Ksar of Ait Ben Haddou
Alberto Pettineo, Giovanni Pancani 123

Thermal efficiency of traditional building materials in diverse climates: The literature review
Doha CHBARI, Oumaima Ait Rami, Kaoutar Ouali 130

A comprehensive life cycle assessment of sustainable reinforcement solutions for expansive soils using natural materials
Ahlam EL MAJID1, Khadija BABA, Latifa OUADIF, Yasmina ED-DARIY 139

Bonding mechanism of hot-pressed green composites using FTIR spectroscopy
Ejazulhaq Rahimi, Ayane Yui, Yuta Yamachi, Yuma Kawasaki 147

Integrating phase change materials for eco-friendly construction: Optimal positioning in building envelopes under Drâa-Tafilalet (Morocco) climate conditions
Azzeddine ELGHOMARI, Amine TILIOUA 156

Development of ceramic membranes based on Draa-Tafilalet clay for Malachite Green retention in water filtration
Mohammed Chrachmy, Ayoub Souileh, Rajae Ghibate, Achraf Mabrouk,
Mohamed Ech-Chykry, Anjoud Harmouzi, Najia El Hamzaoui, Meryem Ben Baaziz,
Hassan Ouallal, Mohamed Azrour 164

Keyword Index

About the Editors

Preface

This volume brings together the latest research findings presented at the GeoME International Scientific Conference on Geosciences and Environmental Management, hosted at the Higher School of Technology in Salé, Morocco, 2025. The conference gathered scholars, researchers, and practitioners from around the world to discuss emerging advances in materials, environmental management, and civil infrastructure, with a particular focus on sustainable and resilient solutions for contemporary challenges in the built environment and heritage conservation.

The contributions collected in this ebook reflect the multidisciplinary spirit of GeoME, covering topics such as innovative construction materials, seismic resilience, sustainable building technologies, and digital approaches to heritage conservation.

We hope that this collection will serve as a valuable resource for academics, professionals, and students interested in the intersection of materials science, environmental sustainability, and civil engineering. By sharing these diverse perspectives, we aim to foster further dialogue, collaboration, and innovation in addressing the pressing issues facing our communities and built environments today.

Committees

Honorary Committee

Mr. Azzedine EL MIDAOUI. Minister of Higher Education and Scientific Research and Innovation
Mr. Mohamed KHALFAOUI. Secretary General of the Department of Higher Education and Scientific Research
Mr. Mohamed RHACHI. President of Mohammed V University in Rabat
Mrs. Lalla Badr Saoud ALAOUI. President of Miftah Essaad Foundation for the Intangible Capital of Morocco
Mrs. Jamila EL ALAMI. Director of National Center for Scientific and Technical Research
Mr. Abderrahman NOUNAH. Director of Higher School of Technology in Salé
Mr. Hassane MAHMOUDI. Director of Mohammadia Engineering School, Rabat, Morocco
Mr. Lahcen BAHI. International Expert in Geoengineering and Sustainable Development

Local Organizing Committee

BABA K. Higher School of Technology of Sale, Morocco
BAHI L. Mohammadia Engineering School, Rabat, Morocco
DOUGHMI K. Mohammadia Engineering School, Rabat, Morocco
EL RHAFFARI Y. Higher School of Technology of Sale, Morocco
KHLIFATI O. The Moroccan School of Engineering, Rabat, Morocco
LAMDOUAR N. Mohammadia Engineering School, Rabat, Morocco
MASROUR I. Mohammadia Engineering School, Rabat, Morocco
SIMOU S. Moroccan School of Engineering Sciences of Rabat, Morocco
OUADIF L. Mohammadia Engineering School, Rabat, Morocco

Organizing Committee

AHATRI M. National School of Applied Sciences of Agadir, Morocco
AKKOURI N. Mohammed VI Polytechnic University, Ben Guerir, Morocco
AMMAR A. Higher School of Technology of Sale, Morocco
AZROUR M. Faculty of Science and Technology of Errachidia, Morocco
BABA K. Higher School of Technology of Sale, Morocco
BAHI A. Mohammadia Engineering School, Rabat, Morocco
BAHI L. Mohammadia Engineering School, Rabat, Morocco
BELHAJ S. National School of Architecture, Marrakech, Morocco
BENRADI F. Higher School of Technology of Sale, Morocco
BERKALOU K. Mohammadia Engineering School, Rabat, Morocco
BOUAJAJ A. National School of Applied Sciences of Al Hoceima, Morocco
BOUGHOU N. Higher Institute of Nursing Professions and Health Techniques in Rabat, Morocco
CHAKIRI S. Ibn Tofail University of Kenitra, Morocco
CHERKAOUI E. Higher School of Technology of Sale, Morocco
ED-DARIY Y. National School of Architecture, Fez, Morocco
EL AKKARY A. Higher School of Technology of Sale, Morocco
EL BAHLOULI T. Rabat School of Mines, Morocco
EL HACHMI D. Faculty of Sciences, Rabat, Morocco
EL JALIL M. H. Higher School of Technology of Sale, Morocco
ELLOUZE A. National Engineering School of Sfax, Tunisia
ELLOUZE S. National Engineering School of Sfax, Tunisia

EL MAJID A. Moroccan School of Engineering Sciences of Rabat, Morocco
EL MENDILI Y. Special School for Public Works, Building, and Industry (ESTP), France
EL RHAFFARI Y. Higher School of Technology of Sale, Morocco
HAJJI A. Higher School of Technology of Sale, Morocco
KHAMAR M. Higher School of Technology of Sale, Morocco
LAAROUSSI N. Higher School of Technology of Sale, Morocco
LAHLOUH I. Higher School of Technology of Sale, Morocco
LAHMILI A. Mohammadia Engineering School, Rabat, Morocco
LAMDOUAR N. Mohammadia Engineering School, Rabat, Morocco
MENZHI M. National Center for Scientific and Technical Research, Rabat, Morocco
MOUHAT O. Higher School of Technology of Sale, Morocco
NOUNAH A. Higher School of Technology of Sale, Morocco
OUADIF L. Mohammadia Engineering School, Rabat, Morocco
RAZZOUK Y. National School of Applied Sciences, El Jadida, Morocco
SEFIANI N. Higher School of Technology of Sale, Morocco
SIMOU S. Moroccan School of Engineering Sciences of Rabat, Morocco
SIROUX M. INSA Strasbourg , France
TILIOUA A. Faculty of Science and Technology of Errachidia, Morocco
ZERROUK L. Higher School of Technology of Sale, Morocco

International Scientific Committee

AALIL I. Ecole Nationale Supérieure d'Arts et Métiers, Meknes, Maroc
AARAB A. High Normal School of Rabat, Morocco
AKKOURI N. Mohammed VI Polytechnic University, Ben Guerir, Morocco
ALAOUI HAFIDI A. Abdelmalek Essaâdi University, Tetouan, , Morocco
AMMAR A. Higher School of Technology of Sale, Morocco
ARIF KAMAL M. School of Architecture & Planning, Amity University, Jaipur, India
AZENHA M. University of Minho, Portugal
AZROUR M. Faculty of Science and Technology of Errachidia, Morocco
BABA K. Higher School of Technology of Sale, Morocco
BAHI A. Mohammadia Engineering School, Rabat, Morocco
BAHI L. Mohammadia Engineering School, Rabat, Morocco
BARBUTA M. Technical University Gheorghe Asachi Iasi, Romania
BOUASSIDA M. National Engineering School of Tunis, Tunisia
BOULANOUAR A. National School of Applied Sciences of Al Hoceima, Morocco
CAMARA A .B. Thiès Polytechnic School, Senegal
CHAKIRI S. Ibn Tofail University of Kenitra, Morocco
CHERKAOUI E. Higher School of Technology of Sale, Morocco
CHERKAOUI M. National Graduate Engineering School – Mines, Rabat, Morocco
CHERRADI T. Mohammadia Engineering School, Rabat, Morocco
CYR M. INSA, Toulouse, France
DERBAL HABAK H. University of Picardy Jules Verne, France
DOBRYNINA A. Institute of the Earth's Crust, Siberian Branch, Russian Academy of Science, Russia
EL AKKARY A. Higher School of Technology of Sale, Morocco
EL BAHLOULI T. Rabat School of Mines, Morocco
EL BRAHMI J. Mohammadia Engineering School, Rabat, Morocco
EL HACHMI D. Faculty of Sciences, Rabat, Morocco
EL HAMMOUMI O. Hassan II University of Casablanca, Morocco
EL HARROUNI K. National School of Architecture, Rabat, Morocco
EL JALIL M. H. Higher School of Technology of Sale, Morocco

ELLOUZE A. National Engineering School of Sfax, Tunisia
ELLOUZE S. National Engineering School of Sfax, Tunisia
EL MENDILI Y. Special School for Public Works, Building, and Industry (ESTP), France
EL RHAFFARI Y. Higher School of Technology of Sale, Morocco
ELAHMADI Z. Mohammadia Engineering School, Rabat, Morocco
ESSAHLAOUI A. Moulay Ismail University of Meknes, Morocco
FAYE C. S. Thiès Polytechnic School, Senegal
FEIZ A. Université d'Evry Paris-Saclay, France
FRIKHA W. National Engineering School of Tunis, Tunisia
GAROUM M. Higher Normal School, Rabat, Morocco
GHORBEL E. University of Cergy-Pontoise, France
GONZÁLEZ F. Faculdade de Arquitectura, Universidade de Lisboa, Portugal
HAJJI A. Higher School of Technology of Sale, Morocco
HAKKOU R. Faculty of Sciences and Techniques, Marrakech, Morocco
HAMED A. National Research Institute of Astronomy and Geophysics, Egypt
KHABBAZI A. Higher School of Technology of Sale, Morocco
KHAMAR L. National School of Applied Sciences of Khouribga, Morocco
KHAMAR M. Higher School of Technology of Sale, Morocco
LAAROUSSI N. Higher School of Technology of Sale, Morocco
LAHLOUH I. Higher School of Technology of Sale, Morocco
LAHMILI A. Mohammadia Engineering School, Rabat, Morocco
LAMDAOUAR N. Mohammadia Engineering School, Rabat, Morocco
MENZHI M. National Center for Scientific and Technical Research, Morocco
MOUHAT O. Higher School of Technology of Sale, Morocco
NABAWY B.S. National Research Centre, Cairo, Egypt
NOUNAH A. Higher School of Technology of Sale, Morocco
OUADIF L. Mohammadia Engineering School, Rabat, Morocco
RAHMOUNI A. Sidi Mohamed Ben Abdellah University, Fez, Morocco
SAFHI A. E. Ostfold University College, Norway
SATWANT S.R California Polytechnic State University, United States
SEFIANI N. Higher School of Technology of Sale, Morocco
SBARTAÏ Z.B. University of Bordeaux, France
SIROUX M. INSA Strasbourg, France
SOUFI A. Mohammadia Engineering School, Rabat, Morocco
TANARHTE M. Faculty of Sciences and Technologies, Mohammedia, Morocco
TAYEH B. A. Islamic University of Gaza, Palestine
TILIOUA A. Faculty of Science and Technology of Errachidia, Morocco
TOKATLI C. Trakya University, Turkey
ZERROUK L. Higher School of Technology of Sale, Morocco

Emerging Research in Materials for Environment, and Civil Infrastructure - GeoME 5.5   Materials Research Forum LLC
Materials Research Proceedings 58 (2026) 1-8   https://doi.org/10.21741/9781644903933-1

Comparative seismic vulnerability of shear walls and moment frames in RC buildings

Sana Afnzar[1,a *], Mohamed Ahatri[1,2,b], Mina Derife[1,c], Abderrahman Atmani[1,d], El Hassan Ait Laasri[1,e]

[1]GMES Laboratory, National School of Applied Sciences, Ibn Zohr University, Agadir, Morocco

[2]University of Mohammed VI Polytechnic (UM6P), Benguerir 43150, Morocco

[a]sana.afnzar@edu.uiz.ac.ma, [b]m.ahatri@uiz.ac.ma, [c]m.derife@uiz.ac.ma,
[d]a.atmani@uiz.ac.ma, [e]e.aitlaasri@uiz.ac.ma

Keywords: Seismic Vulnerability, Pushover Analysis, RC Structure, Bracing Systems

Abstract. This paper seeks to assess and compare the seismic vulnerability of shear walls (SW) and moment-resisting-frames (MRF) as lateral resistance systems in reinforced concrete structures, with application to structures located in the seismic region of Agadir city, Morocco. The main goal of this research is to adapt structural design methodologies to help enhance structural resilience and inform design choices in regions disposed to medium to high-frequency seismic activity. A non-linear static analysis, through the pushover technique, was conducted on two building models, one bridged by shear walls and the other by moment-resisting-frames using performance-based seismic design principles. These models represent a typical mid-rise reinforced concrete structure commonly used by engineers in the region. Therefore, the results show that moment-resistant frame structures have higher seismic vulnerability than shear wall systems, particularly in the X-axis direction. In addition, these findings provide a comprehensive understanding of the trade-offs between stiffness and ductility in seismic design, particularly in urban contexts where architectural flexibility and seismic safety need to be balanced. For example, the estimated probabilities of slight, moderate, and severe structural damage in frame systems were approximately 100%, 98%, and 58%, respectively, compared to only 68%, 34%, and 12% in shear wall systems. The same result was recorded for the Y-axis, with relatively lower overall values.

Introduction

Reinforced concrete structures' response to seismic loads in seismic-prone areas is a subject of great interest to researchers and specialists. The structural system plays a crucial role in a building's performance during earthquakes by resisting horizontal forces. Among the most commonly used structural systems, we find moment-resisting frames (F) and shear walls (SW).

However, choosing an ideal structural system requires taking into consideration: height, use of the building, architectural constraints, and seismic level risk at the construction site. The importance of making the right choice is particularly evident in high or moderate seismic risk regions, Agadir city that situated in southern of Morocco as an example, which suffered a devastating earthquake in 1960, resulting in terrible loss of life and property. This tragic event had a profound impact on the city's history and played a decisive role in the creation of Morocco's first earthquake-resistant building code, RPS 2000 revised RPS 2011[1].

More recently, the region experienced several other seismic events, most notably the Al Haouz earthquake of September 8, 2023, which was the most significant, registering a magnitude of 6,7. Although Agadir was not near the epicenter, the city was strongly shaken. Fortunately, no major damage was reported in the city, conversely, several villages near Al Haouz were severely affected and experienced widespread destruction.

Emerging Research in Materials for Environment, and Civil Infrastructure - GeoME 5.5　　Materials Research Forum LLC
Materials Research Proceedings 58 (2026) 1-8　　　　　　　　　　https://doi.org/10.21741/9781644903933-1

Several researchers have investigated the effect of the bracing system on structural resistance, stiffness, and overall seismic behaviour. For instance, Y. Razzouk et al. [2,3] demonstrated that the choice of an adequate bracing system is necessary for designing a seismic-resistant building, as the effectiveness of different types of bracing varies according to soil conditions and building height, which influence displacement and structural stability under seismic loads. M. Mouhine et al.[4] studied the influence of shear walls in enhancing the seismic behaviour of irregular reinforced concrete structures, reporting a significant improvement in resistance and a reduction in vulnerability. In the context of fragility curve elaboration based on the non-linear analysis method, several studies have been performed, among which we can mention [5,6].

This research aims to study and compare the fragility curves of a reinforced concrete structure braced by walls and another configuration of the same building but braced by moment-resisting-frames. Hence, a nonlinear static analysis was adopted to assess the vulnerability of each system and develop its fragility curves. This analysis allowed for a comparison of the seismic performance of these two systems when subjected to horizontal forces. The results of the study provide recommendations for choosing the most appropriate resistance system, particularly in areas exposed to seismic activity and comprising mid-rise buildings.

Simulation and Computational Methods

Characterization and Numerical Modeling. The building used as a case study in this work is a reinforced concrete residential building located in Agadir, Morocco. This city is considered to be an area prone to strong earthquakes according to the Moroccan RPS code [1].

The building has a rectangular shape, measuring 15 meters in length and 10 meters in width. It is composed of a 2.7 m basement, a 4.2 m ground floor, and three stories, each with a height of 3m. Table 1 shows the structural elements' maximum dimensions and their cross sections in both configurations.

Based on the architectural plan of the building shown in Fig. 1, two structural configurations were designed in conformity with the requirements of the Moroccan seismic design code RPS-2000, updated in RPS-2011. The first configuration shows bracing by walls, while the second is based on a moment resisting frame (column-beam), as shown in Fig. 2. Both variants have been designed to maintain architectural compatibility while satisfying regulatory criteria in terms of regularity, rigidity, and seismic resistance. Non-linear behavior is accounted for by incorporating plastic hinges at beam and column ends, according to ATC-40 [7,8]. The structural material used is C25/30 concrete, with 25 MPa compressive strength.

Nonlinear Pushover Analysis and Fragility Curve Development. In this research, the non-linear static pushover analysis (NLPA) technique is adopted to evaluate the structure's seismic behavior. The study consists of progressively applying lateral loads or displacements until the structures reach their ultimate capacity. The structural capacity is illustrated by the force–displacement curve. Subsequently, the base shear and top displacement are transformed respectively into the equivalent SDOF system's spectral acceleration and displacement. The curves are idealized into their bilinear form following FEMA 273 guidelines [9]. Two key performance points are identified: The yield (D_y,A_y) and ultimate (D_u,A_u) performance points [10]. Based on these points, fragility curves are developed. The later express the probability of a structure reaching a certain damage level depending as a function of a parameter indicating the seismic intensity, which, in this study, is represented by the spectral displacement S_d in this work. The fragility curve is characterized by a cumulative lognormal distribution, characterized by a standard logarithmic deviation β_{dsi} and mean spectral displacement $\overline{Sd}_{ds_i}$.

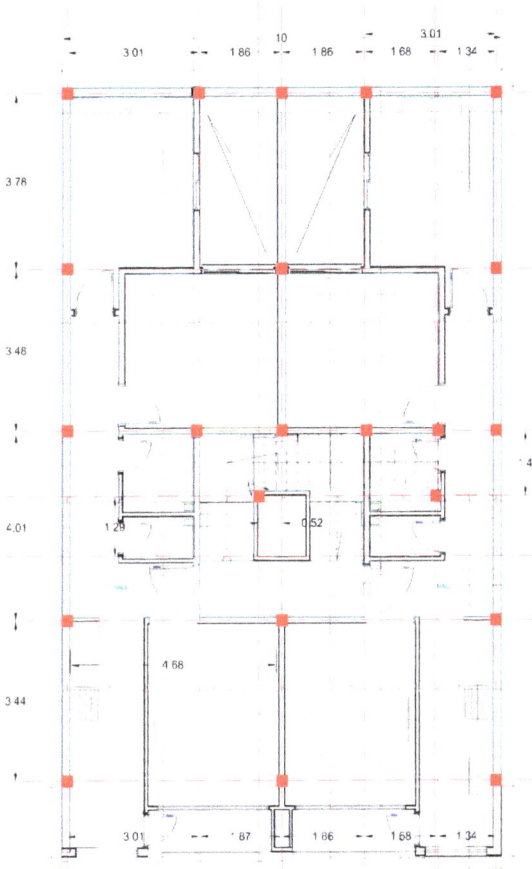

Fig. 1. *Architectural Plan of the Current Floor.*

Table 1. *Maximum Cross-Sectional Dimensions of Structural Elements for Both Configurations.*

Structural Element	Shear Wall Configuration	Moment Frame Configuration
Columns	25×25 [cm]	40×60 [cm]
Beams	20×40 [cm]	30×40 [cm]
Shear Walls (Thickness)	20 [cm]	—

3

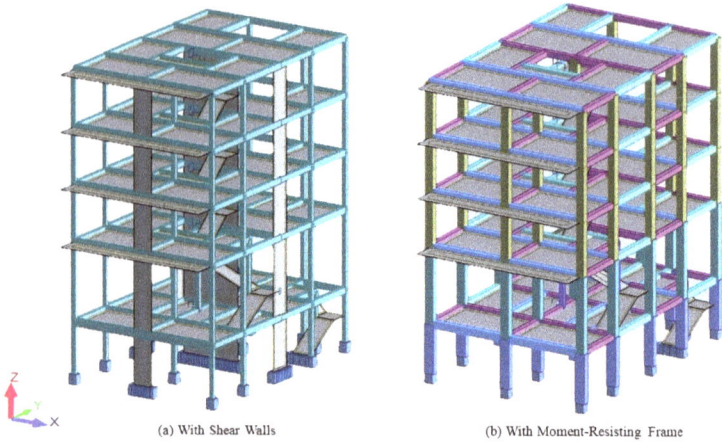

(a) With Shear Walls   (b) With Moment-Resisting Frame

***Fig. 2.** Structural Modelling of the Two Seismic Bracing Configurations.*

The fragility curve for a specified damage state ds_i, is given by the function below [11]:

$$P[ds_i/S_d] = \emptyset \left[\frac{1}{\beta_{ds_i}} ln \left(\frac{S_d}{\overline{Sd}_{ds_i}} \right) \right] \quad (1)$$

Where:

$\emptyset$: Normal cumulative distribution function

Table 2, illustrates the procedure for calculating the thresholds $\overline{Sd}_{ds_i}$ as a function of the structure's Dy and Du displacements [10,12].

***Table 2.** Damage Evaluation Limits.*

Damage State	Threshold Spectral Displacement
Slight	$\overline{Sd}_{ds_1} = 0.7 \times D_y$
Moderate	$\overline{Sd}_{ds_2} = D_y$
Severe	$\overline{Sd}_{ds_3} = D_y + 0.25(D_u - D_y)$
Complete	$\overline{Sd}_{ds_4} = D_u$

Results and discussion

Evaluation of Global Displacements According to RPS 2000 Revised RPS 2011. In this section, the global displacements derived from the linear dynamic analysis are presented and evaluated in conformity with the displacement limits distinct by the Moroccan-Seismic-Design Code RPS-2000, reviewed RPS-2011. The results were computed for both structural configurations: moment-resisting frames and shear walls. These values are then compared with the admissible threshold Δ_g, which is calculated as follows [1] :

$$\Delta_g = 0.004 \times H \quad (2)$$

Emerging Research in Materials for Environment, and Civil Infrastructure - GeoME 5.5   Materials Research Forum LLC
Materials Research Proceedings 58 (2026) 1-8   https://doi.org/10.21741/9781644903933-1

Where:

H: Total structural height.

Table 3 provides a summary of the global displacement values for the two structural systems, as well as the admissible limit according to the RPS 2000 revised RPS 2011. The results showed that, for both configurations, the displacements remained below the admissible threshold, demonstrating that the moment resisting frame system as well as the shear wall systems provide sufficient lateral stiffness to meet the seismic design requirements.

Table 3. *Global Displacement Values for Both Structural Configurations.*

	Structural System	UX [cm]	UY [cm]	Admissible Displacement [cm]
Seismic RPS 2000 revised RPS 2011 Dir. X	Moment Resisting Frame	5.1	1.5	5.28
	Shear Wall	3.9	1.3	5.28
Seismic RPS 2000 revised RPS 2011 Dir. Y	Moment-Resisting Frame	0.7	3.8	5.28
	Shear Wall	2	4.7	5.28

Fragility Curves Development. To analyze the seismic vulnerability for each bracing system configuration, we developed its capacity curves by performing a series of nonlinear static analyses using the pushover method. These curves were then transformed into capacity spectra. To estimate the mean-spectral-displacement ($\overline{Sd}_{ds_i}$) and its standard deviation (β_{dsi}). The main displacement parameters, including failure and ultimate displacements were derived.

Working from this indicator and the analytical expression given in Eq. 1, the probability of specific damage states was determined. Consequently, the fragility curves obtained, which illustrate these probabilities, are shown in Fig. 3. In addition, the damage probabilities calculated by integrating the displacements derived from the RPS 2000 revised RPS 2011 (Table 3) in both orthogonal directions (X and Y) for each configuration are shown in Fig. 4.

The Fig. 3 results highlight the fragility curves of the configurations studied along the X and Y axes. It was observed that all damage cases indicate that the frame structure (F) reflects higher overturning probabilities than the shear wall structure (SW), indicating increased vulnerability when exposed to seismic loading. This trend confirms that moment-resistant frame systems, at the same level of spectral displacement, exhibit early failure and consistently high overturning probabilities in all damage cases. This is evident in the uniform displacement of the fragility curves towards lower displacement values, indicating lower resistance to seismic demand. On the other hand, shear wall systems have curves that are typically shifted towards higher values of displacement, illustrating their greater ability to delay the occurrence of damage. Additionally, Fig. 4 shows the probabilities of reaching each damage state, based on the actual displacements attained for each configuration under the applied seismic load, as per the RPS 2000 revised RPS 2011seismic code requirements. The results demonstrate that moment resisting frames have significantly higher probabilities of exceeding the various damage states than those with shear walls.

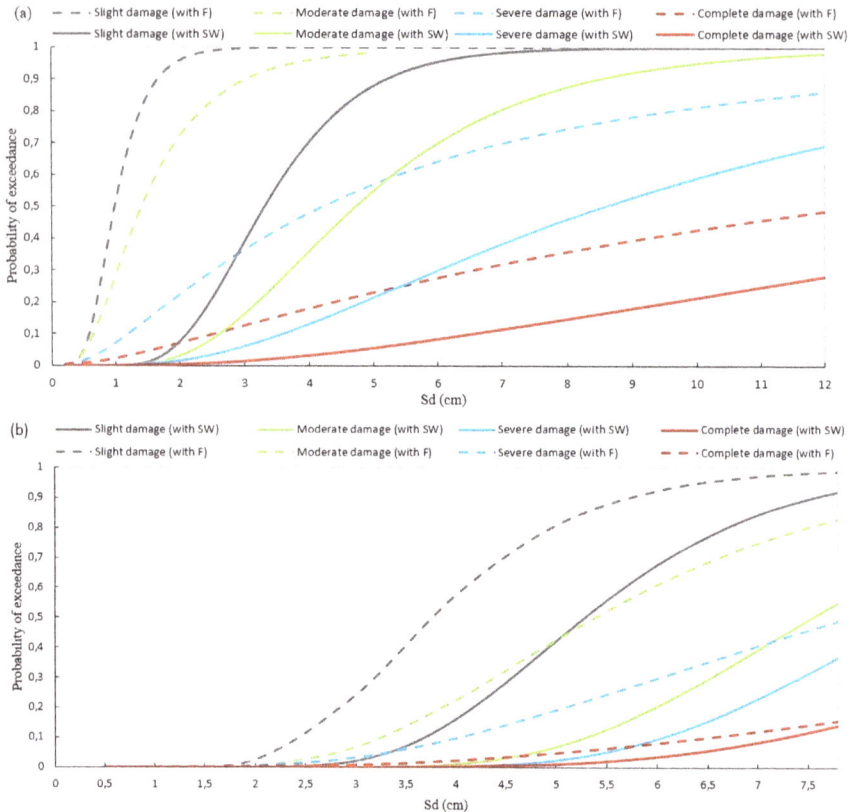

Fig.3. *Computed fragility curves for both directions: (a) X-direction, (b) Y-direction.*

This is particularly evident in the X direction, where the probability of achieving light, moderate and severe damage for rigid frame systems is 100%, 98% and 58%, respectively. In the same direction, shear wall systems have much lower probabilities 68% for light damage, 34% for moderate damage, while for severe damage its only 12%. The same is observed in the Y direction, with lower overall probabilities for both systems. For example, the probability of light damage for the rigid Y-frame system is 51%, whereas it is 34% for the shear wall system. This difference in directionality is due to the natural asymmetry of lateral strength and stiffness, with the X-direction generally being softer and therefore more subject to higher deformation demand under seismic loading.

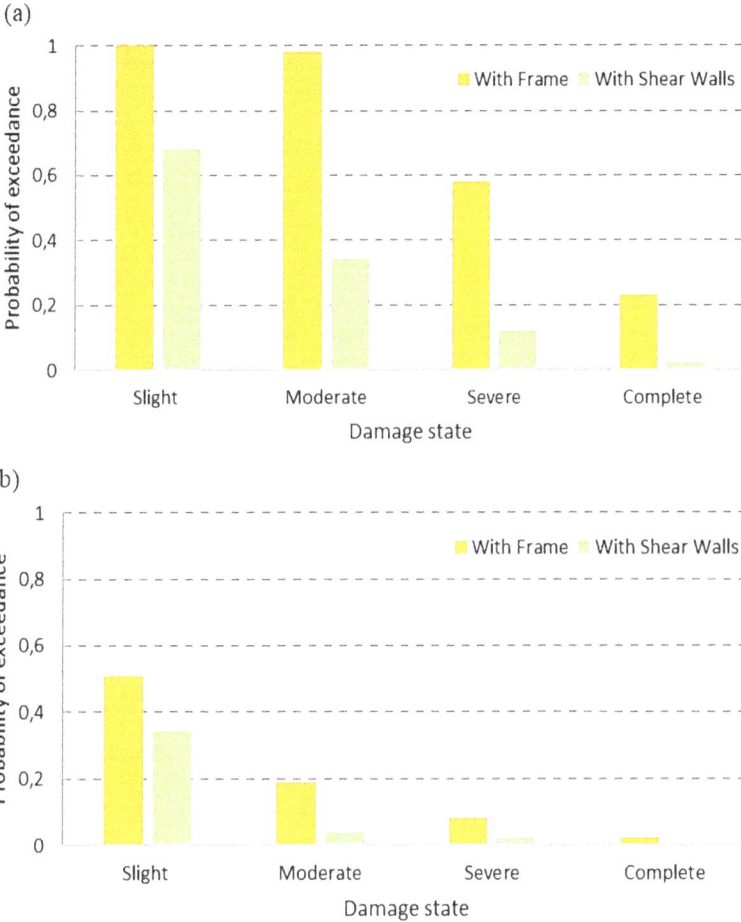

Fig.4. Damage probabilities of studied configurations: (a) X-direction, (b) Y-direction.

Conclusion

A comparison and vulnerability assessment of two structural systems: shear walls and moment resisting frames was carried out under the seismic loading characteristic of Agadir region. The study was conducted using the pushover method, providing capacity curves which were subsequently converted into fragility curves. Based on both fragility analysis and damage probability analysis in the two main directions X and Y, this study provides an overview of the influence of different bracing systems on structural behavior under seismic loads.

In synthesis, the inclusion of shear-walls significantly improves the structure seismic response by enhancing the overall lateral stiffness, the energy dissipation capacity and significantly reducing the probability of reaching the predefined states of damage. These improvements are particularly notable in the direction where the structure demonstrates greater initial flexibility, since shear walls limit lateral deformations and regulate more effectively the evolution of damage.

Therefore, the use of shear walls should be considered as an essential seismic design approach, particularly in regions of high seismic risk, as it results in stronger, more damage resistant and safer structural systems.

Acknowledgments

The National Center for Scientific and Technical Research (CNRST) provided funding for this study through its "PhD Associate Scholarship PASS Program".

References

[1] R.P.S, "Le reglement de construction parasismique," Ministère l'Habitat la Polit. la V. LE, 2011.

[2] Y. Razzouk, M. Ahatri, K. Baba, and A. El Majid, "Optimal Bracing Type of Reinforced Concrete Buildings with Soil-Structure Interaction Taken into Consideration," Civ. Eng. J., vol. 9, no. 6, pp. 1371–1388, 2023. https://doi.org/10.28991/CEJ-2023-09-06-06

[3] Y. Razzouk, M. Ahatri, K. Baba, and A. El Majid, "The Impact of Bracing Type on Seismic Response of the Structure on Soft Soil," Civ. Eng. Archit., vol. 11, no. 5, pp. 2706–2718, 2023. https://doi.org/10.13189/cea.2023.110534

[4] M. Mouhine, M. Derife, S. Aboumdian, and E. Hilali, "Improving Seismic Vulnerability of Irregular Reinforced Concrete Moment-Resisting Frames using Shear Walls," Int. J. Eng. Trans. A Basics, vol. 37, no. 2, pp. 425–438, 2024. https://doi.org/10.5829/ije.2024.37.02b.17

[5] M. Remki, A. Kibboua, D. Benouar, and F. Kehila, "Seismic Fragility Evaluation of Existing RC Frame and URM Buildings in Algeria," Int. J. Civ. Eng., vol. 16, no. 7, pp. 845–856, 2018. https://doi.org/10.1007/s40999-017-0222-7

[6] E. Souki, K. Abdou, and Y. Mehani, "Development of Fragility Curves for Reinforced Concrete", vol. 2, no. 2, pp. 86–96, 2024. https://doi.org/10.23968/2500-0055-2024-9-2-86-96

[7] M. Inel and H. B. Ozmen, "Effects of plastic hinge properties in nonlinear analysis of reinforced concrete buildings," Eng. Struct., vol. 28, no. 11, pp. 1494–1502, 2006. https://doi.org/10.1016/j.engstruct.2006.01.017

[8] R. A. Hakim, M. S. Alama, and S. A. Ashour, "Seismic Assessment of RC Building According to ATC 40, FEMA 356 and FEMA 440," Arab. J. Sci. Eng., vol. 39, no. 11, pp. 7691–7699, Oct. 2014. https://doi.org/10.1007/s13369-014-1395-x

[9] D. Shapiro, C. Rojahn, L. D. Reaveley, J. R. Smith, and U. Morelli, "NEHRP Guidelines and Commentary for the Seismic Rehabilitation of Buildings," Earthq. Spectra, vol. 16, no. 1, pp. 227–239, 2000. https://doi.org/10.1193/1.1586092

[10] S. E. Cherif, A. Chaaraoui, M. Chourak, M. Oualid Mghazli, A. El Omari, and T. M. Ferreira, "Urban Seismic Risk Assessment and Damage Estimation: Case of Rif Buildings (North of Morocco)," Buildings, vol. 12, no. 6, 2022. https://doi.org/10.3390/buildings12060742

[11] A. H. Barbat, L. G. Pujades, and N. Lantada, "Seismic damage evaluation in urban areas using the capacity spectrum method: Application to Barcelona," Soil Dyn. Earthq. Eng., vol. 28, no. 10–11, pp. 851–865, 2008. https://doi.org/10.1016/j.soildyn.2007.10.006

[12] M. Mouhine and E. Hilali, "Seismic vulnerability assessment of RC buildings with setback irregularity," Ain Shams Eng. J., vol. 13, no. 1, p. 101486, 2022. https://doi.org/10.1016/j.asej.2021.05.001

Emerging Research in Materials for Environment, and Civil Infrastructure - GeoME 5.5   Materials Research Forum LLC
Materials Research Proceedings 58 (2026) 9-16   https://doi.org/10.21741/9781644903933-2

Concrete strength prediction: Exploring an ensemble machine learning approach

Younes Alouan[1,2,a] *, Seif-Eddine Cherif[1,b], Badreddine Kchakech[2,c],
Youssef Cherradi[2,d] and Azzouz Kchikach[1,3,e]

[1]Georesources, Geoenvironment and Civil Engineering Laboratory (L3G), Cadi Ayyad University, Morocco

[2]LAMIGEP Research Laboratory, Moroccan School of Engineering Sciences, Morocco

[3]Geology and Sustainable Mining Institute (GSMI), Mohammed VI Polytechnic University, Morocco

[a]y.alouan.ced@uca.ac.ma, [b]s.cherif@uca.ma, [c]b.kchakech@emsi.ma,
[d]y.cherradi@emsi.ma, [e]a.kchikach@uca.ma

Keywords: Concrete, Compressive Strength, Artificial intelligence, Machine Learning Algorithms, Ensemble Learning

Abstract. The mechanical properties of concrete depend on a number of variables, which include the water cement ratio, the properties of materials, and the curing method. Recently, there has been an interest in the use of artificial intelligence (AI) methods. The current investigation utilized a number of Machine Learning (ML) models, including SVR, ANN, RFR, AdaBoost, XGBoost, and GBM, for making predictions of the compressive strength of concrete. The models considered in this investigation could offer accurate values. The data was collected from previous studies, which provided information on 1030 concrete samples. These samples were spread over a wide range of mix proportions and curing ages, thus guaranteeing good generalization capability of the models. The comparison of conventional models with ensembled models showed that there was a marked difference in favor of ensembled models after hyperparameter optimization. The R^2 value of XGBoost was seen to be 0.94 with a MAPE of 10.4% that was almost equal to the ANN model, while the R^2 value was seen to be 0.89 for the SVR.

Introduction

A civil engineering project's lifecycle, from design to long-term maintenance, is a complex and multidimensional process [1]. Machine learning (ML) algorithms are integrated into every stage of this cycle, including planning, building, maintenance, and structural design [2]. Civil engineering has advanced significantly by employing ML, offering sophisticated tools to handle challenging problems at every stage of the project lifecycle. Among these challenges, predicting concrete compressive strength (CS) stands as a critical aspect of structural design and material performance assessment, where machine learning emerges as a powerful predictive tool. ML approaches provide advanced data-driven frameworks for predicting concrete performance through the interpretation of previous experimental data [3]. Numerous studies have applied ML-based models to forecast the properties of concrete in both its fresh and hardened states.[4], where these properties depend on the concrete formulation and the quantity and quality of the constituent materials [5]. CS is the property that engineers most commonly use when designing concrete structures [6]. Concrete CS is typically determined through destructive testing of samples, a procedure requiring significant time and financial resources. To address these challenges, researchers have investigated ML approaches to estimate concrete strength based on mix design input. In 1998, Yeh pioneered the application of machine learning (ML) to predict concrete strength, integrating linear regression with artificial neural networks (ANNs) to assess the compressive strength of high-performance concrete using a dataset comprising 1030 samples. [7].

The existing literature has dedicated itself to the examination of how the CS of various concrete types could be successfully estimated by various types of Machine Learning algorithms. Several sophisticated models of HPC, RAC, SCC, or other nontraditional models of concrete could be successfully estimated by various Machine Learning approaches concerning artificial neuronal networks (ANN), SVM, random forests (RF), Ensembling approaches, or gradient boosting machines (GBM). The conventional models of SVR or ANN could be effectively applied, registering significant progress, thereby establishing an R^2 of 0.83 by SVR, as reported by Xu et al. [8], or an R^2 of 0.93 by ANN by Arslan et al. on a dataset of 1637 patterns [9]. However, Ensembling approaches of bagging or Boosting models frequently demonstrated more accuracy or effectiveness than conventional models. The random forests (RF), a variant of the bagging approach, demonstrated remarkable accuracy, reporting an R^2 of 0.9729 or an MAE of 3.9423 MPa by Mai et al. [10]. Boosting algorithms also demonstrated exceptional accuracy, reporting an R^2 of 0.97 by Wang et al. by employing XGB on a dataset of 228 samples [11] or an R^2 of 0.952 by AdaBoost on a dataset of 1030 samples by Feng et al. [12]. The concrete strength can be estimated by employing the use of Machine Learning models on the premise of the dataset's effectiveness, size, or hyperparameter tuning. The existing research work in this particular domain aims to examine how conventional models of ANN or SVR or Ensembling models of random forests (RFR), AdaBoost (ADB), XGB (XGB), or Gradient Boost (GBM) could be effectively used on the same dataset. It becomes highly significant to strike an efficacy level concerning improving the accuracy or reducing the possibility of fitting on the 'overfitted' hyperparameters, employing GridSearchCV.

Models and methodology

The use of artificial intelligence (AI), more specifically machine learning (ML), has recently emerged as an important area of research in modeling complex materials' responses. In general, there are three types of machine learning methodologies: supervised learning, unsupervised learning, or reinforcement learning. The current research work will be instructed by the supervised learning concept, with focus on regression models which try to forecast the numerical values of variables, like the CS of concrete, from characteristics of its design mixture [3]. Several regression models had been developed by previous studies in an attempt to tackle this predictive problem, with a particular focus on capturing the concrete's CS variation. The models that will be considered in this current research include SVR, ANN, RFR, AdaBoost, XGBoost, and GBM. The models mentioned earlier will be briefly attributed by relevant studies from the literature in Table 1.

Learning the relationship between concrete characteristics and CS is possible through the use of ML models. Based on vast datasets, the model can extract relationships between various parameters that impact compressive strength. Once learned these relationships, they can predict the CS of concrete for new mix formulations. A dataset of 1030 samples, compiled from existing literature, was employed in this work. Each sample contains eight input features corresponding to the concrete mixture parameters (seven of these variables are related to the formulation: the quantity of cement, mineral admixture content, water content, admixture content, fine and coarse aggregate content, all expressed in Kg/m^3 and the 8th input is a variable indicating the crushing day), in addition to these 8 input variables, we have only one target value (CS) (table 2). Providing the model with input values and the target, and that the target is a continuous value, corresponds to what is called supervised learning by regression.

Emerging Research in Materials for Environment, and Civil Infrastructure - GeoME 5.5   Materials Research Forum LLC
Materials Research Proceedings 58 (2026) 9-16   https://doi.org/10.21741/9781644903933-2

Table 1. Overview of studies assessing ML models for strength prediction

ML Models	Number of Input Variables	Dataset Size	R^2 (Test Set)
SVR [8]	7 [C, BF, Fly.A, W, S, CA, FA]	425	0.83
ANN [9]	8 [W, C, FA, CA, BF, Fly.A, S, D]	1637	0.93
RFR [8]	7 [C, BF, Fly.A, W, S, CA, FA]	425	0.93
ADB [8]	7 [C, BF, Fly.A, W, S, CA, FA]	425	0.9
XGB [11]	8 [W/B Ratio, W, FA-Ratio, S, Air-Entraining -Agent, Slump, Air-Content, D]	228	0.97
GBM [13]	9 [C, W, CA, FA, Recycled CA, Fly.A, S, Water Absorption of RA, and Density of RA]	314	0.92

Table 2. Statistical characteristics of the parameters used for model development

Parameter	Abbr	Mean	Min	Max	Std
Cement Content	C	281.2	102.0	540.0	104.5
Slag (Blast Furnace)	BF	73.9	0.0	359.4	86.3
Fly-Ash	Fly.A	54.2	0.0	200.1	64.0
Mixing Water	W	181.6	121.7	247.0	21.3
Superplasticizer admixture	S	6.2	0.0	32.2	6.0
Coarse Aggregate Content	CA	972.9	801.0	1145.0	77.7
Fine Aggregate Content	FA	773.6	594.0	992.6	80.2
Curing Age	D	45.7	1.0	365.0	63.2
Compressive strength	CS	35.8	2.3	82.6	16.7

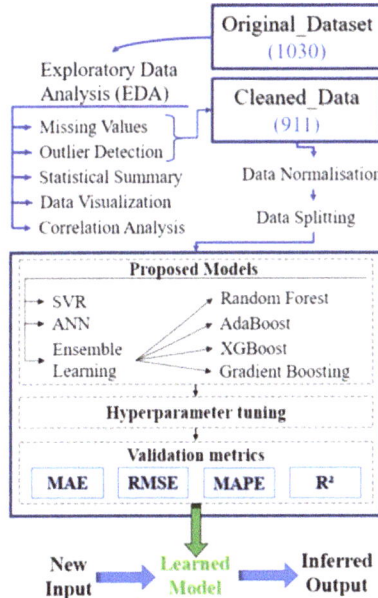

Fig. 1. Research Methodology Process

Emerging Research in Materials for Environment, and Civil Infrastructure - GeoME 5.5   Materials Research Forum LLC
Materials Research Proceedings 58 (2026) 9-16   https://doi.org/10.21741/9781644903933-2

The research process followed in this case comprises a series of steps including data exploratory analysis, preprocessing, development of models with hyperparameters, testing of models, and comparison of models with existing models defined by literature (Fig. 1). Initially, an Exploratory Data Analysis (EDA) was performed on the collected data comprising 1030 samples. There is an investigation into data distribution, pattern development, or the discovery of association among variables. The EDA will help identify various data patterns through graphs created from histograms and scatter plots. Moreover, it will help identify the potentially unbalanced association of data points among variables. For this purpose, a correlation matrix was provided (Fig. 2). Results show that cement quantity showed the highest association value of 0.46 with CS, followed by an association of degree 0.52 with CS, suggesting an increase in cement quantity will strengthen concrete. However, it was discovered that W showed an association value of -0.34 with CS, which shows an inverse association suggesting more quantity of W will deteriorate concrete strength. According to this statement, an opposite association was found among superplasticizer (-0.62), suggesting that more usage of superplasticizer can help decrease the quantity of W.

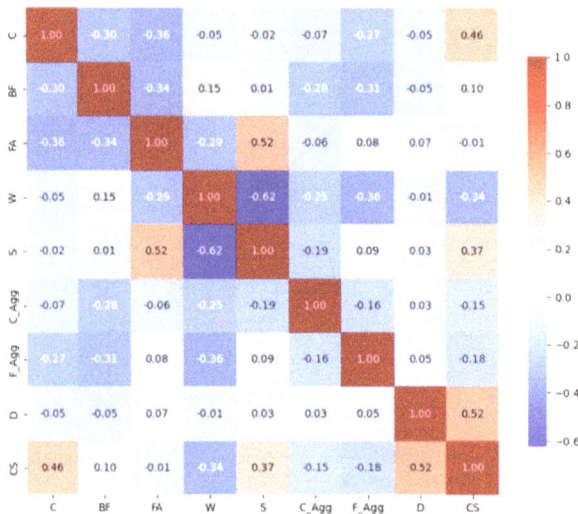

Fig. 2. Correlation Matrix

To make it clearer how all of these parameters correlate with each other, a scatter diagram of (CS) with respect to the W/C ratio was drawn, using age (D) in the form of a blue gradient. This shows an inverse relationship whereby CS reduces with an increase in W/C, thus reconfirming that excessive amounts of water weaken concrete. Moreover, it can be seen that older concrete has higher strength (bluer regions in Fig. 3).

The dataset was then put through a preprocessing step before building the models. The first step was the assessment of missing values in the dataset. The assessment indicated that there was no incomplete data in the dataset. The second step was outlier removal. The removal of outliers was performed in order to ensure that there was no data that would contribute to erroneous predictions by the models. The total number of outliers identified across the dataset was 119, resulting in a dataset of 911 samples from an original 1030. The third step was data normalization. The dataset was then put through data normalization. Data normalization was done with the intention of scaling all the variables equally. The dataset has variables measured in various units. Data normalization

Emerging Research in Materials for Environment, and Civil Infrastructure - GeoME 5.5   Materials Research Forum LLC
Materials Research Proceedings 58 (2026) 9-16   https://doi.org/10.21741/9781644903933-2

avoids dominance of a particular variable in the learning process. Data preprocessing is fundamental in ensuring that machine learning happens successfully. The models rely on proper data. Each of the models' hyperparameters was then tuned by using a grid search with cross-validation to ensure that their predictions could be optimal while not overfitting.

To assess the accuracy of the models, four statistical measures: MSE, MAE, MAPE, and R^2, were used. The four measures provide a comprehensive measure of accuracy of the models in terms of the generalization of the models. The four measures assess how well a proposed or existing model can approximate the actual values of the data. The best models will have an R^2 of almost 1. The models will also ideally have low values of MAE, MSE, or MAPE.

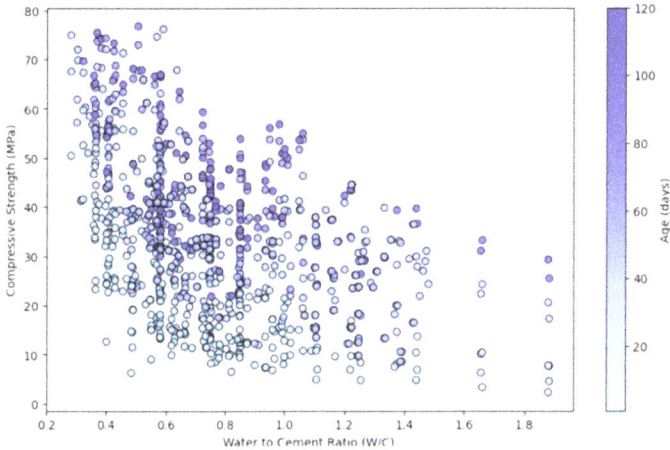

Fig. 3. Compressive Strength Development: Role of Water-Cement Ratio and Age

Results and discussion
The performance of the six algorithms was assessed, and it was observed that the models that performed best after tuning surpassed their performances in the untuned models on all counts (Fig. 4). To begin with, it was seen that the ensemble models of XGBoost, Gradient Boosting, and Random Forest performed better than all others, including an R^2 measure of 0.92 and 0.88 on the test dataset, along with low values of MAE and RMSE. However, the SVR model performed poorly (R^2 measure of 0.65, MAPE of approximately 30.17%), which was sensitive to kernel parameters [8]. The above findings indicate the natural resilience of the ensemble models, specifically the boosting models, in dealing with nonlinear patterns of the variables cement, water, aggregates, and age. As earlier stated, research by Feng et al. found AdaBoost performing well in anticipating the strength of concrete [12].

After hyperparameter optimization, there seems to be an overall improvement in all models. The optimal hyperparameters found by GridSearchCV, which led to this improvement, can be seen in Table 3. After optimization, it was seen that all models showed significant improvement on the testing data, with the largest improvement seen in SVR (+37% in R^2 values) and near perfect training accuracy ($R^2 \approx 0.99$) by Learning. XGB, GBM ($R^2 = 0.94$), followed by an insignificant gain than ADB, RFR, ANN ($R^2 = 0.92$), continued to display the best generalization.

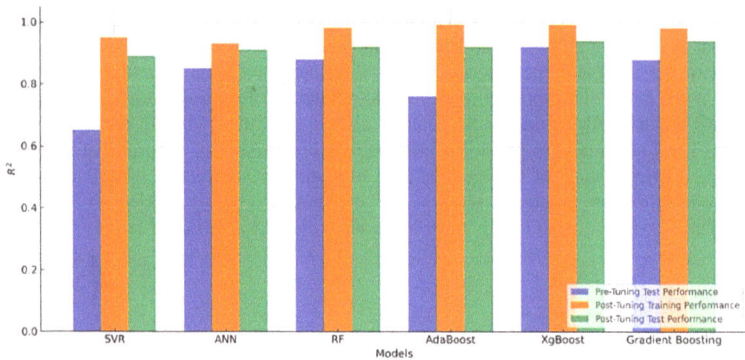

Fig. 4. Comparison of Models Performance Before and After Tuning

Table 3. Optimal Hyperparameters Identified via GridSearchCV

ML Models	Hyperparameters	
SVR	C = 100; Epsilon = 0.1	Gamma = auto; Kernel = rbf
ANN	Hidden layer = (18,); Activation = tanh; Regularization = 0.0001	Solver = SGD; Learning rate = 0.001; Epoch = 1000
RFR	Nbr-estimators = 200; Depth-max = 20; Features-max = log2	Samples-split-min = 2; Samples-leaf-min = 1
ADB	Estimator = DT (depth=10); Nbr-estimators = 100	Learning-rate-value = 0.1; Loss = linear
XGB	Nbr-estimators = 300; max-depth = 5; Gradient-update-factor = 0.1	Min-child-weight =3; sub-sample = 0.8; Colsample-bytree = 0.8
GBM	Nbr-estimators = 300; Depth- max = 4; Learning-rate-value = 0.1	Samples-split-min=10;subsample = 0.9; Samples-leaf-min = 1

Observation of particular interest is the trade-off between complexity and accuracy. Even though the models with the best accuracy belong to the boosted tree category, they tend to be complicated in nature. Overfitting was identified from the comparison of training sets with testing sets of various models, as seen in Fig. 4. For example, AdaBoost showed an R^2 of 0.99 on the training set but only 0.92 on the testing set. The difference in mean absolute error was substantial, with an increase from 0.43 MPa on the training set to 3.39 MPa on the testing set, thereby implying that it has memorized the data. Similar observations on Random Forest indicated an R^2 of 0.98 on the training sets compared to 0.92 on the testing sets. The degree of overfitting was moderate. This can be attributed to an increase in the complexity of models owing to parameters like a large number of estimators (RF, 200 estimators; AdaBoost, 100 estimators). However, it is pertinent to note that the ANN showed a remarkably stable level of generalization. The difference in ANN accuracy on both training (R^2 of 0.93) and testing sets (R^2 of 0.92) was merely ~1%. XGBoost (seen in Fig. 4), on comparison, showed similar levels of accuracy on both training (R^2 of 0.99) and testing sets (R^2 of 0.94), which could be attributed to the combined effect of regularization techniques of subsampling and learning rate. An almost similar level of difference was seen in Gradient Boost.

The findings from this research show the importance of hyperparameter tuning in the use of ML models in the prediction of concrete CS. The difference in the accuracy of models during hyperparameter tuning shows that models' inherent settings do not provide optimal accuracy. The findings of this research show that Ensemble models perform better than single models (figure 5 below) in this case of SVR and ANN models. The effectiveness of Ensemble models in this research can basically be attributed to how the models combine outputs of multiple models, thus resulting in an optimal combination of bias and variance. Hence less prone to both underfitting and overfitting than single models. Even though ANN models can provide an optimal solution by hyperparameter tuning (This involves selecting the right network structure, i.e., configuration of hidden units & layers).

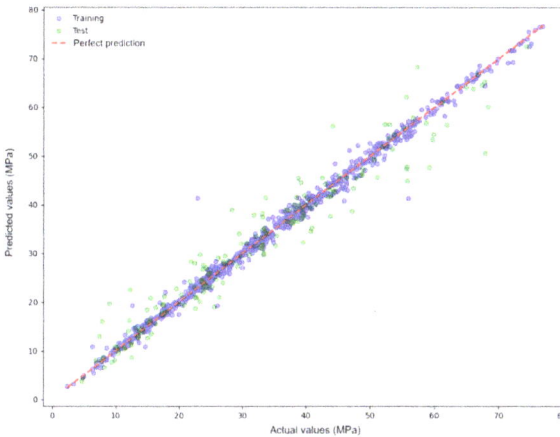

Fig. 5. XGBoost Model Performance: Actual Versus Predicted Results

Conclusion

This investigation conducted a comprehensive comparison of six ML techniques to estimate the CS of concrete mixtures, emphasizing the critical role of hyperparameter optimization. Our results show that boosted tree ensembles, particularly XGBoost and Gradient Boosting, once finely tuned, deliver the best predictive accuracy for concrete CS, outperformed other approaches in terms of accuracy and generalization, achieving test R^2 scores up to 0.94 after tuning. These models benefited from inherent regularization strategies, effectively mitigating overfitting while capturing complex nonlinear interactions between input variables such as C, BF, Fly.A, W, S, CA, FA and curing age. Although models like SVR and ANN initially showed limited performance under default configurations, their predictive power improved substantially with tailored hyperparameter tuning, with SVR showing the highest relative gain in R^2 (+37%). Notably, the ANN model exhibited exceptional stability, with minimal discrepancy between training and test performance, making it a promising single model alternative when properly configured. The findings underscore that model selection alone is insufficient; careful tuning of architecture and parameters is indispensable for unlocking each algorithm's full potential. Ultimately, this work reinforces the superiority of ensemble methods in balancing bias-variance trade-offs and provides practical guidance for the deployment of machine learning models in concrete mix design optimization.

References

[1] D. M. Frangopol and M. Soliman, "Life-cycle of structural systems: recent achievements and future directions," *Structure and Infrastructure Engineering*, vol. 12, no. 1, 2016. https://doi.org/10.1080/15732479.2014.999794

[2] H. Sun, H. V. Burton, and H. Huang, "Machine learning applications for building structural design and performance assessment: State-of-the-art review," Jan. 01, 2021, *Elsevier Ltd.* https://doi.org/10.1016/j.jobe.2020.101816

[3] M. Mohtasham Moein *et al.*, "Predictive models for concrete properties using machine learning and deep learning approaches: A review," Jan. 01, 2023, *Elsevier Ltd.* https://doi.org/10.1016/j.jobe.2022.105444

[4] I. Nunez and M. L. Nehdi, "Machine learning prediction of carbonation depth in recycled aggregate concrete incorporating SCMs," *Constr Build Mater*, vol. 287, Jun. 2021. https://doi.org/10.1016/j.conbuildmat.2021.123027

[5] G. N. Kumar and G. V. V. Satyanarayana, "Strength and durability properties of quaternary blended high strength concrete," in *E3S Web of Conferences*, EDP Sciences, Jun. 2023. https://doi.org/10.1051/e3sconf/202339101205

[6] J. Huang, M. M. S. Sabri, D. V. Ulrikh, M. Ahmad, and K. A. M. Alsaffar, "Predicting the Compressive Strength of the Cement-Fly Ash–Slag Ternary Concrete Using the Firefly Algorithm (FA) and Random Forest (RF) Hybrid Machine-Learning Method," *Materials*, vol. 15, no. 12, Jun. 2022. https://doi.org/10.3390/ma15124193

[7] I. C. Yeh, "Modeling of strength of high-performance concrete using artificial neural networks," *Cem Concr Res*, vol. 28, no. 12, pp. 1797–1808, 1998. https://doi.org/10.1016/S0008-8846(98)00165-3

[8] Y. Xu *et al.*, "Computation of high-performance concrete compressive strength using standalone and ensembled machine learning techniques," *Materials*, vol. 14, no. 22, Nov. 2021. https://doi.org/10.3390/ma14227034

[9] A. Qayyum Khan, H. Ahmad Awan, M. Rasul, Z. Ahmad Siddiqi, and A. Pimanmas, "Optimized artificial neural network model for accurate prediction of compressive strength of normal and high strength concrete," *Cleaner Materials*, vol. 10, Dec. 2023. https://doi.org/10.1016/j.clema.2023.100211

[10] H. V. T. Mai, T. A. Nguyen, H. B. Ly, and V. Q. Tran, "Prediction Compressive Strength of Concrete Containing GGBFS using Random Forest Model," *Advances in Civil Engineering*, vol. 2021, 2021. https://doi.org/10.1155/2021/6671448

[11] W. Wang, Y. Zhong, G. Liao, Q. Ding, T. Zhang, and X. Li, "Prediction of Compressive Strength of Concrete Specimens Based on Interpretable Machine Learning," *Materials*, vol. 17, no. 15, Aug. 2024. https://doi.org/10.3390/ma17153661

[12] D. C. Feng *et al.*, "Machine learning-based compressive strength prediction for concrete: An adaptive boosting approach," *Constr Build Mater*, vol. 230, Jan. 2020. https://doi.org/10.1016/j.conbuildmat.2019.117000

[13] A. H. A. Ahmed, W. Jin, and M. A. H. Ali, "Prediction of compressive strength of recycled concrete using gradient boosting models," *Ain Shams Engineering Journal*, vol. 15, no. 9, p. 102975, Sep. 2024. https://doi.org/10.1016/J.ASEJ.2024.102975

Emerging Research in Materials for Environment, and Civil Infrastructure - GeoME 5.5   Materials Research Forum LLC
Materials Research Proceedings 58 (2026) 17-23   https://doi.org/10.21741/9781644903933-3

Numerical study and buckling behavior of web-posts in stainless-steel cellular beams

Imane Bachguar[1,a*], Ouadia Mouhat[1,b], Rabee Shamass[2,c],
Fatima El Mennaouy[1,d]

[1]Mohammed V University, Mohammadia Engineering School, Civil Engineering and Environment Laboratory (LGCE), Rabat, Morocco

[2]Department of Civil and Environmental Engineering, Brunel University London, UK

[a]imanebachguar@gmail.com, [b]ouadie.mouhat@gmail.com, [c]shamassr@lsbu.ac.uk, [d]fatimaelmennaouiy@gmail.com

Keywords: Stainless Steel, Carbon Steel, Cellular Beams, Buckling, FEM

Abstract. Owing to their structural efficiency, cellular steel beams are increasingly used in construction. These beams are characterised by a unique shape that lightens the structure without compromising its strength. In terms of mechanical performance, the presence of these openings causes a concentration of stresses around them, which leads to a redistribution of forces within the beam. In this work, we will explore the use of stainless steel cellular beams in order to highlight the unique aspects of their mechanical behaviour, which varies from that of carbon steel. The most significant difference lies in the configuration of the stress-strain curve. While carbon steel is characterised by linear-elastic stress-strain behaviour, followed by a plateau before work hardening begins, stainless steel has no clearly defined yield strength and exhibits non-linear response and significant work hardening. In order to perform our finite element simulations, stainless steel cellular beams with different characteristics and material properties were used. The objective is to study the structural behaviour of these beams and to analyse the effect of the beam geometry and mechanical characteristics on the overall structural response. In this regard, a mechanical model was adjusted and validated using the results of numerical analyses, thus enabling the phenomenon of buckling failure of the web element to be described.

Introduction

Cellular steel beams have become an essential tool in the quest to design smarter, more efficient steel structures. The presence of these openings allows all building services, including plumbing, electrical systems and ventilation ducts, to be integrated into the depth of the beam. These cellular beams are also characterised by their high degree of design flexibility and can be used for a variety of purposes, whether for commercial buildings, industrial facilities or even large-span structures. Thanks to their structural performance, aesthetics and flexibility of use, cellular beams are increasingly being used in a wide range of projects.

They were initially developed to meet the demands of long spans in industrial and commercial buildings. Their design not only reduces the structural weight and therefore material costs but also facilitates the integration of technical systems (heating, ventilation, electricity) through the openings built into the beam itself. With advancements in computational tools and manufacturing methods, it is now possible to design composite cellular beams suitable for spans of up to 18 meters, without compromising safety or structural durability.

The manufacturing method for these cellular beams is based on a relatively simple yet precise process. It generally begins with the production of an I-section steel beam, which is then cut lengthwise using thermal cutting techniques. The two resulting halves are offset and welded together in such a way as to produce circular openings in the web of the beam. The key dimensional

Emerging Research in Materials for Environment, and Civil Infrastructure - GeoME 5.5   Materials Research Forum LLC
Materials Research Proceedings 58 (2026) 17-23   https://doi.org/10.21741/9781644903933-3

characteristics of cellular beams are: H which represents the cellular beam height, do the opening height, t_w and t_f represents the web and flange thicknesses, s_o the spacing from edge to edge separating two successive openings and s the length between the centres of the diameters of the openings.

Generally, several types of failures can affect cellular beams. Vierendeel bending (VB) occurs around the openings when the vertical segments of the web bend under loading. Global bending (BF) refers to excessive deformation of the entire beam under heavy loads. Shear failures (SF) are also observed, especially near the openings, where transverse forces are concentrated. Web or post-buckling (WPB) can occur if the areas between the openings lack sufficient stiffness. There is also the risk of lateral-torsional buckling (LTB), which affects the beam's lateral stability. Lastly, rupture of weld joints may occur at the joints between the two halves of the beam, especially if the welds are poorly executed or overstressed.

The behavior of cellular steel beams has been examined by Several researchers. Lawson et al.[1] suggested a calculation method that considers the asymmetry of the cross-section of the cellular beam. This asymmetry has a considerable impact on the moments between the openings. For the buckling of web columns, equations have been proposed, based on the 'strut' model. Erdal and Saka [2] studied the non-linear behavior of steel cellular beams. Twelve I-beams subjected to a point load applied at the center of their span were tested. They found that the most common failure mode is Web Post Buckling (WPB). Panedpojaman et al. [3] suggested a calculation equation to accurately estimate the shear strength of symmetrical or asymmetrical, non-composite cellular beams. To verify and confirm the accuracy of the developed finite element model, several experimental tests were employed for this purpose. A total of 390 simulations were carried out, taking account of geometric imperfections. The proposed formulas allow a more reliable and cost-effective design of cellular beams. Beams with openings of various shapes were used by Tsavdaridis and D'Mello [4] to examine their structural behaviour. A total of seven beams were studied. This study aims to identify and analyse the different failure modes to which a cellular beam may be subjected. Then, using the finite element method, fourteen numerical models were developed in order to compare the results obtained experimentally with those acquired numerically.

In addition to these researchers, others, such as Grilo et al. [5], have used cellular beams of different shapes to conduct several experimental tests with the aim of studying the failure mode of these beams. Their studies have shown that the failure of most beams is due to web-post buckling WPB. In some cases, failure is caused by a combination of WPB and LTB (lateral-torsional buckling), demonstrating the complex behaviour of these structures under load.

In publication SCI 100 [6], the first internationally developed design method was presented. The purpose of this approach is to analyse the buckling of cellular beams. Subsequently, the AISC used the same design method [7]. Lawson et al. [1] introduced a new method for assessing the web post shear buckling resistance. This method is founded on the strut analogy and the equivalent buckling curve in Eurocode 3 [8]. This approach was developed by Panedpojaman et al. [3], taking into account the boundary conditions of the struts.

In this study, we will use stainless steel as the material, which is characterised by its mechanical properties, namely its high ductility and resistance to high temperatures [9].

In this study, different geometric configurations of stainless steel cellular beams were used to examine their structural behaviour. Finally, using the finite element method, several simulations were carried out to ensure and guarantee the reliability of the results.

Numerical model development and validation
Using Abaqus software, a complete beam model was established. The aim was to study and then evaluate the buckling of cellular beams. A complete model was used rather than a single-column model due to uncertainties regarding the boundary conditions to be applied in the single-column

Emerging Research in Materials for Environment, and Civil Infrastructure - GeoME 5.5   Materials Research Forum LLC
Materials Research Proceedings 58 (2026) 17-23   https://doi.org/10.21741/9781644903933-3

model, which did not allow the overall behaviour of the beam to be accurately reproduced. In our study, we relied on a finite element model developed by Shamass and Guarracino [10] using ABAQUS software. This model proved to be particularly reliable for simulating the behaviour of steel cellular beams. It demonstrated notable accuracy in predicting vertical shear resistance as well as in analyzing failure mechanisms in the web of the beam. To ensure the model's reliability, its results were compared with those of experimental tests executed on simply supported cellular beams under point loads.These tests, performed by Grilo et al. [5] and Tsavdaridis and D'Mello [4], provided a solid benchmark for validating the model's accuracy.To date, no experimental studies have been realized on stainless steel cellular beams. For this reason, and to carry out our parametric study, we relied on a previously validated numerical model.

In this study, we will use a single web model and not the complete specimen. The choice of a single web-post is mainly due to the results obtained by comparing the experimental results and those obtained in our numerical study carried out on single web-post models and on the complete specimen. From the results obtained, it can be seen that the percentage difference is small in the case of the single web post. These results indicate that the single web post produces more accurate estimates.

Stainless steel stress-strain performance
When subjected to stress, stainless steel reacts differently from carbon steel. The stress-strain curve shows the greatest difference. Until it reaches a clearly defined yield point, carbon steel exhibits linear elastic behavior, exhibiting minimal strain hardening. In contrast, stainless steel, has a more rounded curve, no clearly defined yield point, high ductility and high progressive hardening. The behavior of stainless steel is modeled numerically using Rasmussen's two-phase stress-strain approach [11], which provides an accurate representation of its non-linear response.

$$\varepsilon = \frac{\sigma}{E} + 0.002\left(\frac{\sigma}{\sigma_{0.2}}\right)^n \text{ for } 0 \le \sigma \le \sigma_{0.2} \tag{1}$$

$$\varepsilon = \varepsilon_{0.2} + \frac{\sigma - \sigma_{0.2}}{E_2} + \varepsilon_{up}{}^*\left(\frac{\sigma - \sigma_{0.2}}{\sigma_u - \sigma_{0.2}}\right)^m \text{ for } \sigma_{0.2} \le \sigma \le \sigma_u \tag{2}$$

With:

$$\varepsilon_{up}{}^* = \varepsilon_u - \varepsilon_{0.2} - \frac{\sigma_u - \sigma_{0.2}}{E_2} \tag{3}$$

$$\varepsilon_u = \min\left(1 - \frac{\sigma_{0.2}}{\sigma_u}, A\right) \tag{4}$$

$$m = \left(3.5\frac{\sigma_{0.2}}{\sigma_u}\right) + 1 \tag{5}$$

$$E_2 = \frac{E}{1 + 0.002\frac{n}{e}} \tag{6}$$

$$e = \frac{\sigma_{0.2}}{E} \tag{7}$$

Where:
σ: The uniaxial stress;
m and n : Strain hardening coefficients;
E: The modulus of elasticity;
$\varepsilon_{0.2}$: The 0.2% strain corresponding to $\sigma_{0.2}$;
A: elongation parameter;
ε: The uniaxial strain;
E_2: The stress-strain curve slope at $\varepsilon_{0.2}$;
ε_u: The ultimate tensile strain;

Parametric study

Through this parametric investigation, the impact of web-post geometry on the buckling shear load will be examined and better understood.

The following parameters characterize the geometry of web post:

$$e = \frac{d_0}{2} - \sqrt{(\frac{d_0}{2})^2 - (\frac{s-d_0}{2})^2} \qquad (8)$$

$$H = d + \frac{d_0}{2} - e \qquad (9)$$

$$s_0 = s - d_0 \qquad (10)$$

Numerical modelling was conducted using the finite element approach to examine the response of beams to lateral bending. A total of 1,000 cellular beam models were used. This method makes it possible to evaluate the impact of various parameters on the overall behaviour and stability of the beam.

This parametric study was performed by modifying three parameters: opening, spacing and slenderness ratio.

The values assigned to the s/d_o ratio in this study are 1.1, 1.2, 1.3, and 1.4;

The ratio d_o/d is equal to 0.8, 0.9, 1.0, 1.1 and 1.2.

Influence of Material Grade and Opening Geometry on Web-Post Shear Buckling Capacity

This section focuses on analyzing the effect of the geometric ratios d_o/t_w and s/d_o on the buckling shear performance of austenitic stainless steel members. For various values of d_o/t_w (33.07, 37.20, 41.33, 45.46 and 49.60), Fig. 1 displays the change in critical buckling shear load according to the spacing ratio.

Austenitic stainless steels 1.4301, 1.4435 and 1.4541, which are frequently used for their mechanical properties and corrosion resistance, will be the focus of our study. In our case, the deformation is localized around the opening in the beam web, this indicate that there is local buckling. In other words, the concentration of stresses around the opening shows that this zone is a critical point in the design.

In this section, we focus on the impact of the s/d_o and d_o/t_w ratios. These impacts are observed on the shear load due to buckling in elements made from austenitic stainless steels. The diagrams below Fig. 1 illustrate the shear buckling load for stainless steel cellular beams. The analysis indicates that the shear buckling load increases with increasing s/d_o ratio and also increases with decreasing d_o/t_w ratio.

Results and Discussion

Fig. 1 shows that as the s/d_o ratio increases, the buckling shear load also increases. Meaning that the greater the distance between the openings, the greater the beam's resistance to web post buckling. In addition, the buckling shear load increases as the d_o/t_w ratio decreases. This means that a cellular beam with larger openings has a less rigid web column, making it more vulnerable to buckling.

Table 1 shows that grade 1.4435 austenitic steel offers the best mechanical performance compared with the other grades studied. From this result, we can clearly see that the choice of material has a direct impact on the structural capacity of cellular beams. Hence, geometric optimization is not the only criterion to be taken into account to guarantee structural stability. These results show that resistance to shear buckling varies significantly according to the beam's geometric parameters and the grade of stainless steel used. Grade 1.4435 emerged as the best performer, making this material particularly suitable for structures subjected to high loads.

Emerging Research in Materials for Environment, and Civil Infrastructure - GeoME 5.5   Materials Research Forum LLC
Materials Research Proceedings 58 (2026) 17-23   https://doi.org/10.21741/9781644903933-3

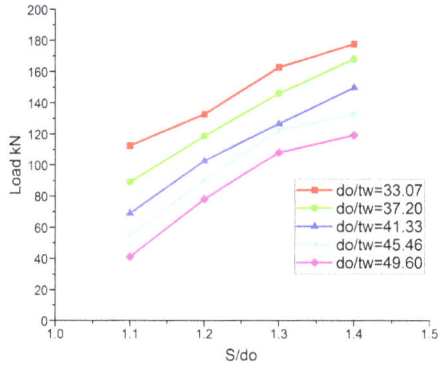

(a)

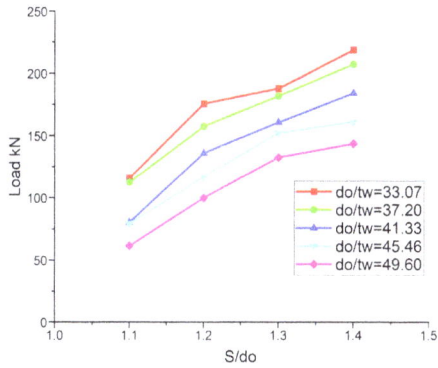

(b)

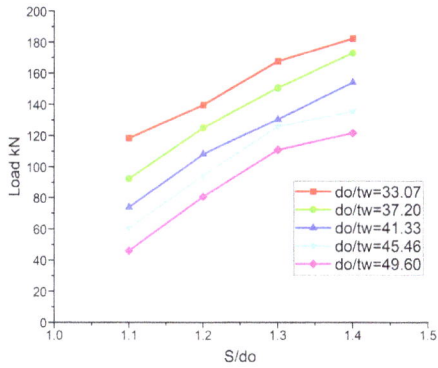

(c)

Fig. 1: Shear buckling load as a function of s/d₀ and d₀/t_w ratios for grades: (a) 1.4301,
(b) 1.4435 and (c) 1.4541

Table 1: Impact of the stainless steel grade on the critical buckling load

Grade	Maximum load [KN]	Material ductility
1.4301	177.37	Average
1.4435	218.43	Excellent
1.4541	182.25	Average

Conclusion

To assess the performance of stainless steel cellular beams with varying geometries, several numerical methods were used. To this end, a set of finite element models was established and then validated using experimental tests carried out on comparable carbon steel beams.

Subsequently, a wide range of simulations were executed to examine the structural response of these cellular beams, taking into account their different shapes and mechanical properties. To anticipate web post buckling (WPB), a simplified mechanical model was formulated and subsequently validated based on the numerical simulation results to propose a rigorous and reliable design method. The study showed that the choice of material used and the geometric characteristics strongly influence the shear buckling resistance of cellular beams.

References

[1] R.M. Lawson, J. Lim, S.J. Hicks, W.I. Simms, Design of composite asymmetric cellular beams and beams with large web openings, Journal of Constructional Steel Research. 62 (2006) 614–629. https://doi.org/10.1016/j.jcsr.2005.09.012

[2] F. Erdal, M.P. Saka, Ultimate load carrying capacity of optimally designed steel cellular beams, Journal of Constructional Steel Research. 80 (2013) 355–368. https://doi.org/10.1016/j.jcsr.2012.10.007

[3] P. Panedpojaman, T. Thepchatri, S. Limkatanyu, Novel design equations for shear strength of local web-post buckling in cellular beams, Thin-Walled Struct. 76 (2014) 92–104.

[4] K.D. Tsavdaridis, C. D'Mello, Web buckling study of the behaviour and strength of perforated steel beams with different novel web opening shapes, Journal of Constructional Steel Research. 67 (2011) 1605–1620. https://doi.org/10.1016/j.tws.2013.11.007

[5] L.F. Grilo, R.H. Fakury, G. de Souza Veríssimo, Design procedure for the web-post buckling of steel cellular beams, Journal of Constructional Steel Research. 148 (2018) 525–541. https://doi.org/10.1016/j.jcsr.2018.06.020

[6] J. Ward, Design of Composite and Non-Composite Cellular Beams, The Steel Construction Institute, SCI Publication, 1990.

[7] S. Fares, J. Coulson, D. Dinehart, Steel Design Guide 31: Castellated and Cellular Beam Design, AISC, 2016.

[8] CEN, Eurocode 3: Design of Steel Structures, Part 1-1 "General Rules and Rules for Buildings", CEN, Brussels: European Committee for Standardization, 2005.

[9] B. Rossi, Discussion on the use of stainless steel in constructions in view of sustainability, Thin-Walled Struct. 83 (2014) 182–189. https://doi.org/10.1016/j.tws.2014.01.021

[10]   R. Shamass, F. Guarracino, Numerical and analytical analyses of high-strength steel cellular beams: A discerning approach, Journal of Constructional Steel Research. 166 (2020) 105911. https://doi.org/10.1016/j.jcsr.2019.105911

[11]   K.J. Rasmussen, Full-range stress–strain curves for stainless steel alloys, Journal of Constructional Steel Research. 59 (2003) 47–61. https://doi.org/10.1016/S0143-974X(02)00018-4

Emerging Research in Materials for Environment, and Civil Infrastructure - GeoME 5.5   Materials Research Forum LLC
Materials Research Proceedings 58 (2026) 24-31   https://doi.org/10.21741/9781644903933-4

Optimization of concrete durability using artificial intelligence: Modeling and prediction of eco-concrete performance with fly ash

Kaoutar BAZZAR[1,a*], Salma CHRIT[1,b], Adil HAFIDI ALAOUI[2,c]

[1]Moroccan School of Engineering Sciences, SMARTILab Laboratory, Rabat, Morocco

[2]University Abdelmalek Essaadi, Faculty of Sciences and Technology, Laboratory of Mechanical and Civil Engineering, Research Team in Materials and Structural Mechanics, Tangier, Morocco

[a]k.bazzar@emsi.ma, [b]s.chrit@emsi.ma, [c]a.hafidi@uae.ma

Keywords: Sustainable Concrete, Fly Ash, Mechanical Strength, Artificial Intelligence, Machine Learning

Abstract. Dealing with the waste and by-products of industry is a major environmental and economic issue. The effort of optimization of materials, especially concrete, is of great importance in construction. Cement production contributes a large share of global CO_2 emissions. Using fly ash to partially replace cement can create eco-friendly concrete which is one of the potential sustainable strategies. This valorization tactic results in lower quantities of fly ash disposed of in dump. In addition, it also reduces the amount of cement consumed, that too without altering the mechanical performance (which is satisfactory). In recent years, artificial intelligence (AI) has proved to be a strong ally helping improve and predict the performance of eco-concrete. One may estimate early-age compressive strength through machine learning algorithms using different parameters like fly ash content or particle size distribution. Instead, deep learning techniques can be used to forecast long-term durability with large experimental and in-situ datasets for characteristics such as ettringite formation and porosity. The approach proposes to study the influence of fineness of fly ash, optimize the key factors affecting the mechanical behavior, and investigate the reasons behind expansion and cracking. All in all, these methods help in making the construction industry sustainable.

Introduction

Concrete is the workhorse of modern construction; However, the cement manufacturing is a primal cause of CO_2 emissions. As proven by the study [17], using fly ash instead of cement mitigates waste and emission in an effective manner which do not contribute to industrial waste and environmental footprint. These properties will depend on the fly ash fineness, C_3A content, porosity, ettringite formation, etc. AI is a power tool that can easily model and optimize these variables to better assess the compressive strength and durability as well as lessen the experimentation. This study aims to create a concrete mix that is mechanically sound and environmentally friendly using AI-based techniques which look at the impact of fly-ash content [16], particles fineness, water-binder ratio and the conditions of curing.

Literature review

Towards More Sustainable Concrete. Challenges and Alternatives. Although concrete is a structurally efficient material, it is among the most polluting materials. Supplementary materials (like fly ash) can aid in achieving circular-economy goals [2]., with mechanical performance and limiting environmental damage [1]. Utilizing fly ash in the mix is consistent with circular economy practices [2] and improves durability and strength [3].

Characteristics of Fly Ash and its Impact on Concrete Performance. Fly ash that results from anthracite or bituminous coal combustion is known as type F. It has high SiO_2 and Al_2O_3 contents and hence it is strongly pozzolanic in nature. Class F fly ash is created using bituminous coal. It is rich in silica and alumina to give strong pozzolanic reactivity [4]. The fineness of material is primarily responsible for its reactivity impact on water demand and early strength. The presence of excess unburnt carbon can cause pH alteration and speed up the corrosion of steel which affects durability.

Pathological Organic Mechanisms and Resistance in HVFA Concrete. Durability problems in eco-concretes occur due to mechanisms like DEF (delayed ettringite formation) [5,6], crystallization pressure [7], and water adsorption [8], causing swelling, cracking and the microstructure weakening [9]. The risks underscore the importance of the optimized water/binder ratio, ash grading, C_3A content and curing conditions.

Old methods versus computerized methods. These enhancements are generally subject to reproducibility and cost problems. Well, AI and ML allow for predictive mix optimization without the need for empirical trials [10]. Thus, the mechanical performance of composite materials can be predicted using models like Random Forest and XGBoost, and neural networks [11, 12]). According to [13] deep learning can further simulate microstructural mechanisms like DEF and can result in predictive, adaptive, and sustainable concrete design.

Issues

High-volume fly ash (HVFA) concrete encounters technical issues, especially the limited pozzolanic reactivity of the fly ash and its variable properties resulting in low early-age strength. Incorrect handling of ASR and delayed ettringite formation to carbonation issues not making durable structures. Due to their expense and ignorance of complex interactions, traditional mix designs are rarely used. Artificial Intelligence [14] is an effective alternative. Therefore, it can assess performance by machine learning. It can also specify critical parameters. Finally, it can also optimize HVFA concrete mix design.

What intelligent AI-based approaches can help to overcome the technical limitations of high fly ash concrete in consideration of complex multi-parametric formulations?

Study objectives

Due to low early-age strength, complex formulation and uncertain long-term durability, HVFA suffer from various problems. Consequently, this study proposes an artificial intelligence-based modeling and optimization approach.

The specific objectives are to. Assess how important parameters influence compressive strength, such as constituents (fly ash content, grading, water/binder ratio), curing time, etc.Study the relations between your Experimental Variables to find the most influencing of the behavior of the HVFA. Utilize various machine learning models to compare and predict strength: Random Forest, XGBoost, MLP. Use R^2, RMSE, and MAE to assess the model performance and find the best method.

Methodology

General structure of the approach.The approach adopted in this study combines experimental analysis and artificial intelligence modeling. It is based on a set of data compiled from tests carried out on concrete containing different proportions of fly ash, together with information on material characteristics, processing conditions and mechanical performance results.

The data was compiled manually [15], homogenized and cleaned [16]. The aim is to model the compressive strength of HVFA concrete from several input variables, using machine learning algorithms, and to identify the most influential variables as well as experimental procedures or protocols for analyzing the importance of mechanical strength.

Database description. The values for fly ash %, W/B ratio, curing time and strength were collated and standardized [15]. works had been compiled, manually and with care, into one dataset. The works covered include that of [17, 18] amongst many others. The data utilized in this study

Emerging Research in Materials for Environment, and Civil Infrastructure - GeoME 5.5   Materials Research Forum LLC
Materials Research Proceedings 58 (2026) 24-31   https://doi.org/10.21741/9781644903933-4

were standardized with respect to units, formulation, and curing conditions used by the algorithms. The dataset includes the following variables.

Table 1. Study criteria

Features	Explanation	Type
% Fly Ash	Fly ash percentage replacing cement.	Quantitative (continuous)
W/B Ratio	Water/binder ratio of the mixture.	Quantitative (continuous)
Time	Cure time (1D, 3D, 7D, 28D), expressed in days.	Qualitative ordinal
Particle Size	Average fly ash particle size (in μm).	Quantitative (continuous)
Density (%)	Relative density of concrete.	Quantitative (continuous)
Compressive Strength	Concrete compressive strength (MPa) - *Target variable.*	Quantitative (continuous)
Tensile Strength	Tensile strength (MPa).	Quantitative (partially missing)
Flexural Strength	Flexural strength (MPa).	Quantitative (partially missing)
Elastic Modulus	Modulus of elasticity (MPa).	Quantitative (partially missing)
Mechanical / Chemical method	Activation technique used (e.g. grinding, low C3A cement, etc.).	Qualitative (indicative)

The data were cleaned, and only observations with values for the target variable (compressive strength) were retained. The **"Time"** column was transformed into a numerical variable for use in regression algorithms (e.g. 1J → 1, 3J → 3, etc.).

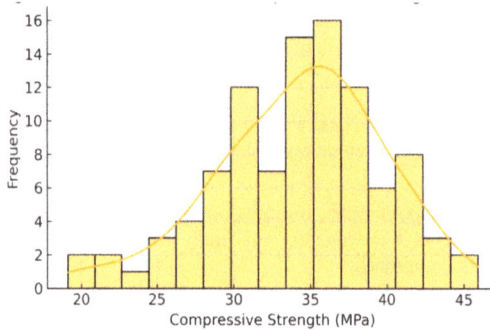

Fig.1. Distribution of compressive strength measured in fly ash concrete samples.

The density curve shows the concentration of data around the central values, revealing a moderate dispersion and slight asymmetry.

Emerging Research in Materials for Environment, and Civil Infrastructure - GeoME 5.5   Materials Research Forum LLC
Materials Research Proceedings 58 (2026) 24-31   https://doi.org/10.21741/9781644903933-4

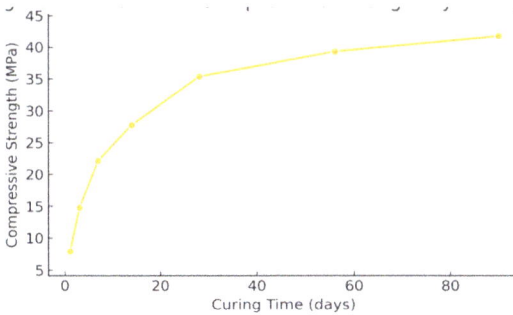

Fig.2. Influence of curing time on compressive strength. A marked progression is observed between 1 and 28 days, confirming the major influence of hydration kinetics on the mechanical development of HVFA concrete.

Exploratory analysis and correlation
Prior to any modeling, a correlation analysis was performed between numerical variables to identify significant relationships with compressive strength. The correlation matrix revealed in particular:

Among the most notable observations was a marked negative correlation between water/binder ratio (W/B) and compressive strength. This outcome is consistent with previous research [17], in particular the studies by [17]. Studies show that too much water causes cement paste to become more porous., thus reducing matrix density and mechanical strength. On the other hand, a low E/L ratio enhances the compactness of the resultant material and facilitates the formation of CSH which contributes to strength gain.

Figure 3 shows the correlation matrix, where variables with strong positive or negative correlations are visually identified by intense hues. This visualization also confirms the importance of curing time and fly ash content, which show significant links with concrete strength.

Data preparation
Before using machine learning, the data set was cleaned for model reliability. The first experimental data contained many columns that were either incomplete or did not matter to the rest of my model. Thus, I removed them and kept only the important ones. Categorical value time (like "1J", "3J", "7J", "28J") which was numeric for modeling. We also ensured that the rows with missing values in the important features as well as the target variable (compressive strength) were deleted from the dataset.

The dataset was divided into 80% for training and 20% for testing to facilitate accurate model evaluation. Also, compare it with previous studies. Normalization was applied finally for the MLP model that is sensitive to scale. Tree based models like Random Forest, XGBoost didn't require that. This preparation step ensured efficient and valid modeling.

Choice of machine learning algorithms
In order to model the complex non-linear relationships in HVFA concrete data, the three algorithms Random Forest, XGBoost and MLP were selected. To create random forests, randomness is added to the bootsrapping of previous classification trees. XGBoost is squashed but requires less memory. Yet fast and accurate gradient boosting machine that learn using regularization. The MLP employs several layers of neurons that are linked to fit deeper non-linearities. This model needs normalization of data. MLP also requires tuning of Hyperparameter

otherwise it may overfit. These models predict the compressive strength to be efficient by using complete data set.

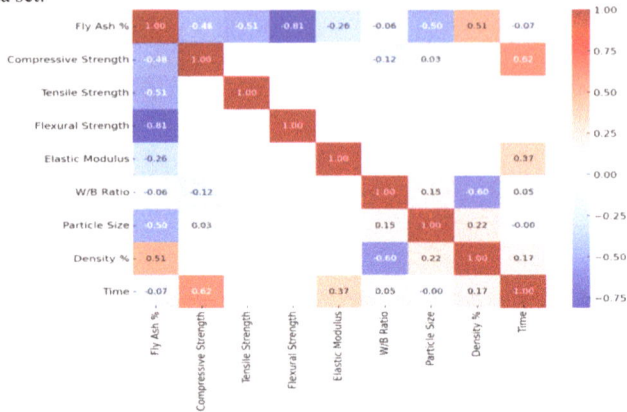

Fig.3. Correlation matrix between quantitative variables. Significant correlations, notably between compressive strength, water/binder ratio and curing time, justify their inclusion in predictive models.

Performance evaluation

The evaluation of machine learning models is done to check their reliability and predictions. To evaluate the prediction quality of the different machine learning methods and the dataset, a total of three measurements will be used in this paper, which are R squared measures the variance that we can explain, RMSE tells us more for big errors, and MAE tells us what our average absolute error is. The value nearer to 1, the better. The average size of the errors in a set of predictions is measured by RMSE. Still, it does so by assigning a relatively high weight to large errors. The MAE indicates the average absolute error in a set of predictions, in the other hand. It provides a better measure of accuracy compared to RMSE. The explained variance, sensitivity to outliers, and overall prediction power parameters, when combined, provide a useful value.

Variable importance analysis

Tree-based algorithms, such as Random Forest and XGBoost, can assess the relative significance of input variables in predicting the outpu . The average reduction in impurity of the decision nodes or how much the accuracy changes on a change of the variable. This method optimizes concrete mix design by highlighting the key parameters namely fly ash content, water/binder ratio and particle fineness. Moreover, it presents an effective means of reducing unnecessary testing. The findings are consistent with physical and chemical laws. They offer useful guidance based on experimental data for the development of high-performance green concrete.

Results and interpretation

Model comparison. Three algorithms Random Forest, XGBoost, and Multilayer Perceptron (MLP), were compared using R2, RMSE, and MAE metrics in order to assess the robustness of the suggested method. With the greatest result (R2 = 0.907, RMSE = 2.666 MPa, MAE = 2.065 MPa), the Random Forest model successfully captured the non-linear nature of the dataset. Although it demonstrated a little higher sensitivity to data fluctuation, XGBoost provided comparable accuracy (R2 = 0.89), and its speed and regularization ability made it a strong alternative. The MLP model's poorer performance was probably caused by its complicated tuning requirements and sparse data. All things considered, tree-based models worked best for predicting

Emerging Research in Materials for Environment, and Civil Infrastructure - GeoME 5.5   Materials Research Forum LLC
Materials Research Proceedings 58 (2026) 24-31   https://doi.org/10.21741/9781644903933-4

HVFA concrete, providing dependable and understandable outcomes that aid in data-driven choices for sustainable material design.

Table 2. Model comparison

Model	R^2	RMSE (MPa)	MAE (MPa)
Random Forest	0.907	2.666	2.065
XGBoost	0.890	3.010	2.490
MLP (Neural Net)	~0.82	~3.65	~2.75

Interpretations. The Random Forest Algorithm's experimental dataset can precisely estimate the high fly ash concrete compressive strength (HVFA). Using input variables such as fly ash, water/binder ratio, particle size, curing time. Random Forest achieved the highest accuracy (R2 = 0.907) of all the models that were tested. This indicates over 90% of the strength variance can be explained by the input variables. The prediction errors were contained with a RMSE of 2.666 MPa and a MAE of 2.065 MPa, showing good robustness of model.

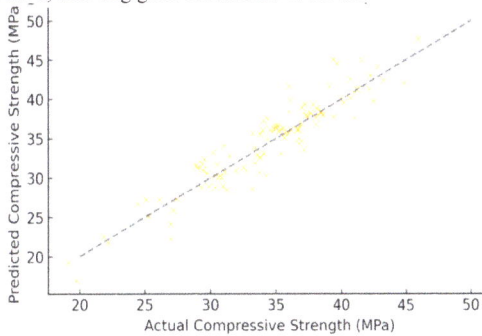

Fig. 4. Scatter plot comparing Random Forest model predictions with actual compressive strength values.

If the points cluster around the diagonal, it indicates a good one-to-one correspondence between predicted values and experimental values. As shown in the Figure 4, predicted values are compared to experimental values and the Random Forest model was able to reproduce measured strength results. The most significant variable affecting 3DPC hardening is curing time, followed by water/binder ratio and fly ash content, as indicated through a variable importance analysis (Figure 5). This 3DPC analysis relates well to previous literature [17, 18], indicating the result will be influenced by pozzolanic reaction kinetics. The water/binder ratio is negatively correlated to compressive strength as well, which aligns with earlier studies on microstructural densification [19]. Our AI-based experimental approach can be executed rapidly and is considerably more flexible than traditional use. This study, which builds on existing modelling work attributed to [20] applies correlation, variable influence, and visualization to derive data-driven recommendations helpful for improving HVFA concrete compositions.

Conclusion

The AI framework presented can effectively predict HVFA performance and can be extended to other binders. When the parameters were optimized, artificial intelligence and machine learning models were able to successfully predict the performance of Portland cement concrete. XGBoost and MLP delivered steady performance, just slightly below that.

The approach we are proposing is consistent with the ones by [17], in addition to being able to have the capability to determine influential variables automatically and reduce physical testing and help in the sustainable mix design. They're definitely going to look at other binders as well as long-term durability in future extensions.

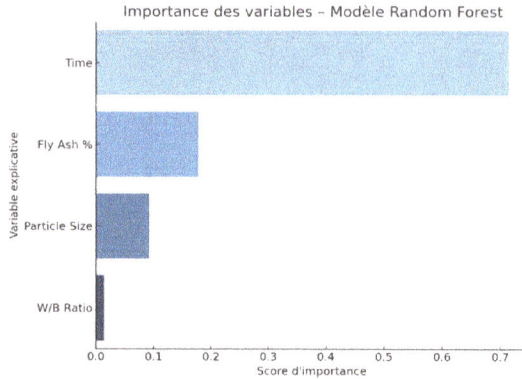

Fig.5. Relative importance of explanatory variables in the Random Forest model. Curing time, water/binder ratio and fly ash content appear to be the most important determinants of compressive strength.

In addition, physical activation through mechanical grinding [21], the use of low-C_3A cement and compaction during curing were noted to be useful in increasing early strength and later durability. While they are not modeled directly, these qualitative factors provide further clarification on the performance differences between formulations and support integrated optimization of materials and processes.

References

[1] B. Han, K. Zhang, X. Yu, E. Kwon, Review: Artificial-intelligence-led revolution of construction materials, Mater. Des. 225 (2023) 111527. https://doi.org/10.1016/j.matdes.2023.111527

[2] P.O. Awoyera, A. Adesina, Sustainable strategies for concrete infrastructure preservation, Infrastructures 10(4) (2025) 99. https://doi.org/10.3390/infrastructures10040099

[3] B.A. Salami, O. Adeyemi, M. Hassan, Multi-objective optimization of sustainable concrete containing fly ash, Buildings 12(7) (2022) 948. https://doi.org/10.3390/buildings12070948

[4] G.L. Golewski, Impact of fly ash composition on pozzolanic activity, Cem. Concr. Res. 45(1) (2022) 126–134. https://doi.org

[5] D. Hunter, Ettringite-induced swelling in soils: State-of-the-art, Appl. Mech. Rev. 48(10) (1995) 659–673. https://doi.org

[6] H.T. Nguyen, Q.T. Bui, T.P. Vo, Prediction of compressive strength of high-volume fly ash concrete using ML models, Constr. Build. Mater. 283 (2021) 122757. https://doi.org/10.1016/j.conbuildmat.2021.122757

[7] G.W. Scherer, Crystallization in pores: Thermodynamic mechanisms of expansion, J. Am. Ceram. Soc. 82(8) (1999) 1957–1968. https://doi.org

[8] R. Siddique, Performance of fly ash in concrete: strength, durability and sustainability, Constr. Build. Mater. 18(3) (2004) 201–208. https://doi.org

[9] J. Stark, K. Bollmann, Delayed ettringite formation in concrete, Cem. Concr. Res. 30(1) (2000) 95–104. https://doi.org

[10] S. Marelli, et al., From concrete mixture to structural design—a holistic optimization, Data-Cent. Eng. 5 (2024) e6. https://doi.org

[11] W. Ha, J. Kim, S. Lee, Influence of unburned carbon in fly ash on hydration, ACI Mater. J. 102(5) (2005) 341–348. https://doi.org

[12] S. Rajakarunakaran, A.M. Kumar, V. Ramasamy, Estimation of concrete strength with Random Forest regression, J. Build. Eng. 45 (2022) 103498. https://doi.org/10.1016/j.jobe.2021.103498

[13] M. Kazemi, S. Mirjalili, AI-driven mix design optimization in sustainable concrete, Eng. Appl. Artif. Intell. 133 (2024) 108099. https://doi.org/10.1016/j.engappai.2024.108099

[14] H.R. Sobuz, M.M. Rahman, M.S. Islam, Durability forecasting of eco-concrete with fly ash using deep learning, Case Stud. Constr. Mater. 20 (2024) e02763. https://doi.org

[15] Z.J. Schaffer, Machine learning-driven optimization of binary concrete mix designs, Ph.D. Thesis, Pennsylvania State University (2024). https://doi.org

[16] P.G. Asteris, et al., Physics-informed modeling of splitting tensile strength of recycled aggregate concrete, Sci. Rep. 15 (2025) 11803. https://doi.org

[17] K. Bazzar, F. Hafiane, A. Hafidi Alaoui, The early-age strength development of substituted mortars with a high-class F fly ash content, Civil Eng. J. 7(8) (2021) 1378–1388. https://doi.org

[18] K. Bazzar, M.R. Bouatiaoui, A. Hafidi Alaoui, Performance approach to the durability of high-volume fly ash concrete, Int. J. Eng. Sci. Innov. Technol. 2(2) (2013) 1–11. https://doi.org

[19] I. Ebrahimi, et al., Enhancing urban resilience: The role of modern concrete materials and AI, Preprint, ResearchGate (2024). https://doi.org

[20] K. Bazzar, A. Hafidi Alaoui, A study on strength properties of concrete with replacement of Low C3A Cement by fly ash, J. Mater. Sci. Eng. B 11(1–3) (2021) 8–15. https://doi.org/10.17265/2161-6221/2021.1-3.002

[21] Y. Su, J. Jiang, L. Wang, Effect of wet-grinding concrete waste slurry on early strength development, Cem. Concr. Compos. 135 (2023) 104812. https://doi.org/10.1016/j.cemconcomp.2023.104812

Emerging Research in Materials for Environment, and Civil Infrastructure - GeoME 5.5   Materials Research Forum LLC
Materials Research Proceedings 58 (2026) 32-39   https://doi.org/10.21741/9781644903933-5

Comparative study of the mechanical and hydro-thermal behavior of earth blocs stabilized by cement or plaster and pozzolan

Imane Daya[1,a] *, Jamila Elbrahmi[1,b], Nouzha Lamdaour[1,c] and Toufik Cherradi[1,d]

[1]Civil engineering, Mohammadia School of Engineering, Mohammed V University in Rabat, Rabat, Morocco

[a]imanedaya.ensmr@gmail.com, [b]elbrahmi@emi.ac.ma, [c]tcherradi@gmail.com, [d]nlamdouar@gmail.com

Keywords: Binder, Pozzolan, Earth Blocs, Mechanical Resistance, Hydrothermal Behavior

Abstract. The building and construction sector is among the most significant contributors to global warming. Indeed, modern construction materials have a major impact on the environment, from their extraction to their installation. Therefore, there is a need to develop new materials that are more durable and better suited to contemporary challenges. In this context, earth materials appear as a sustainable, economical, and environmentally friendly alternative. However, their mechanical strength is relatively low, and their sensitivity to water hinders their development. Thus, the main objective of this study is to assess the effect of using mineral additives, particularly natural pozzolan, as well as the influence of varying the type of binder on the mechanical and hydrothermal performance of earth blocks. In this study, two types of binders were used, namely plaster and cement, at a proportion of 10% by weight relative to the dry mix. In addition, four pozzolan contents (5%, 10%, 15%, and 20%) were incorporated based on the binder weight. After detailing the steps for preparing the samples to be tested in the laboratory according to the selected formulations and percentages, a physico-chemical characterization of the soil used in this study was carried out in order to classify it. This characterization includes the particle size analysis by sieving and sedimentation, plasticity assessment using Atterberg limits, evaluation of the clay fraction via the methylene blue test, and estimation of organic matter content through a calcination test. The research then focused on analyzing the mechanical, thermal, and hydraulic performance of the stabilized earth blocks. To this end, the samples were tested for uniaxial compressive strength and three-point bending to assess their mechanical behavior, thermal conductivity using the steady-state asymmetric hot plane method, and durability through total water absorption testing. Experimental results indicate that blocks treated with plaster exhibit better mechanical performance compared to those treated with cement. Moreover, plaster-stabilized blocks showed lower total water absorption than cement- stabilized blocks throughout the entire immersion period, regardless of pozzolan content. However, plaster is less resistant to water-related degradation: after 96 hours of immersion, cracks and disintegration of the material were observed. The study also showed that compressive strength increases slightly with the addition of pozzolan, up to a certain limit. Beyond 10% pozzolan content, strength decreases. In contrast, flexural strength decreases as pozzolan content increases. On the other hand, the incorporation of pozzolan leads to a reduction in thermal conductivity, and consequently, an improvement in thermal performance, due to increased porosity in the clay matrix

Introduction

The construction sector is responsible for consuming 30% to 40% of global energy, generating about 20% of greenhouse gas emissions, and utilizing a significant amount of natural resources, and their extraction has a considerable impact on biodiversity [1, 2]. Therefore, it is increasingly crucial to develop new materials that are more sustainable and better adapted to the modern

Emerging Research in Materials for Environment, and Civil Infrastructure - GeoME 5.5   Materials Research Forum LLC
Materials Research Proceedings 58 (2026) 32-39   https://doi.org/10.21741/9781644903933-5

context. This brings attention back to earth as one of the most environmentally, economically, and socially responsible materials [3].

In Morocco, the modernization of raw earth began in the early 1960s [4]. This modernization has been accompanied by intensive studies to produce current standards and regulations. It is in this perspective that Morocco established, in 2013, the earthquake-resistant code for earth construction. This regulation contains two parts: the first part is a collection of prescriptions and good practices in terms of seismic protection, developed to assist craftsmen. The second part is the seismic regulation of earth constructions, intended for architects, engineers, and craftsmen involved in construction [5].

The materials used in the construction of traditional building are natural; they are extracted directly from the immediate vicinity. And they are characterized by their diversity [6].

The soil is locally sourced, and its transformation is inexpensive. Additionally, these constructions exhibit exceptional thermal performance, contributing to reduced energy consumption during both construction and operation phases. However, despite its many advantages, earth is sensitive to weather and is characterized by moderate mechanical resistance, which may cause stability problems. In addition, earth structures are very vulnerable to earthquakes due to their fragility and low tensile strength [7]. Stabilizers can be incorporated to enhance the various properties of earth materials. In this context, soil stabilization can be defined as a treatment aimed at improving the mechanical behavior and durability of earth materials by modifying their structure and properties through the addition of binders, fibers, additives, waste materials, etc. [8].

Materials and methods

The materials used. Sol: In this study, we use soil from the Settat region in central Morocco because it is readily available. Tests to characterize this soil were carried out at the Labotest laboratory in Kénitra, and the results are presented in table 1.

Table 1 Characteristics of the soil used

Type of test	Designation	Results
Granularity (sieving and sedimentation)	Gravel	12 [%]
	Sand	28 [%]
	Silt	42 [%]
	Clay	18 [%]
Water content	W	0,7 [%]
Methylene blue value	VBS	6,7
Organic matter content	C	1,45 [%]
Atterberg limit	IP	14,26

The soil is loamy and inorganic, with low plasticity and a notable presence of active clay.

Plaster: It is a plaster marketed by the "Compagnie Marocaine de Plâtre et d'Enduit", from the Safi region. Table 2 summarizes the characteristics of the plaster used in this study.

Table 2 Technical characteristics of the plaster used in this study

Characteristics	Values
Density	1150 [Kg/m^3]
Compressive strength	8 [N/mm^3]
Elasticity modulus	2,327 [N/mm^3]
Conductivity	0,39 [W/$m.K^{-1}$]

Cement: The cement used in this study is CPJ35 Portland cement from "Asment Temara".

Pozzolan: In this study, we used a natural pozzolan of volcanic origin extracted from the Hebri Mountain deposit, located about 20 km from Ifrane, in the Middle Atlas region of Morocco. The identification tests for this pozzolan are shown in Table 3. [9]

Table 3 Physical properties of Habri pozzolan

Designation	Value
Class test	0/5
Dry Density of the pozzolan stone [kg/m3]	1845
Bulk density [Kg/m3]	1150
Specific density [kg/m3]	2360
Porosity [%]	51,27
Absorption factor [%]	14,14

Experimental technique. Preparation of the sample: The specimens are made using the classic technique, namely: Soil preparation, stabilizer dosing (the different formulations used in this study is presented in Table 4), mixing, molding and drying.

Table 4 The different formulations used in this study

Mineral addition Pozzolan [%] (In relation to the weight of binders)	Binder (cement or plaster) [%] (In relation to the dry mix)
5	
10	10
15	
20	

Description test. Simple compression test: This test involves subjecting a sample to uniaxial compression until failure, in order to determine its compressive strength in dry conditions. The compressive strength of the blocks is given by the following formula:

$$\sigma = \frac{F}{S}$$ (Eq. 1)

σ: The compressive strength in MPa
F: Maximum load supported by the sample in N
S: Average surface area of test faces in mm^2

Description test. Three point bending test: This test aims to determine the deformation of a sample under a transverse load.

Emerging Research in Materials for Environment, and Civil Infrastructure - GeoME 5.5   Materials Research Forum LLC
Materials Research Proceedings 58 (2026) 32-39   https://doi.org/10.21741/9781644903933-5

Description test. Water absorption test: This test involves placing a sample in a container with the water level maintained at 1 cm above the bottom of the sample. The increase in mass is then measured after 1, 2, 3, and 4 days. The total absorption is defined by the following relationship:

$$A\% = \frac{P_h - P_s}{P_s}$$ (Eq. 2)

P_h : Increase in weight at time t
P_s : Dry block weight

Description test. Asymmetrical hot plane method in steady state: The asymmetrical hot plate method in steady state allows one to identify the thermal conductivity of a material. The estimate of the thermal conductivity of the sample is given by the following relationship.

$$\lambda_1 = \frac{e_1}{S*(T_0 - T_1)} * \left[\frac{U}{R} - \frac{\lambda_2 * S}{e_2} * (T_0 - T_2)\right]$$ (Eq. 3)

Where e_1 and s are the thickness and the surface of the sample, $\lambda_2 = 0{,}04$ W.m^{-1}.K^{-1} and $e_2 = $ 10mm are successively the thermal conductivity and the thickness of the insulating foam; U is the voltage imposed across the terminals of the heating element with electrical resistance R=40Ω.

Results
Mechanical Behavior. The experimental program of this study will be divided into three phases:
- Study the influence of the nature of the hydraulic binder
- Analyze the effect of pozzolan dosage
- Understand the influence of curing conditions

Results for compressive strength and flexural strength as a function of hydraulic binder type, pozzolan dosage and curing time are shown in Fig. 1 and Fig. 2, respectively

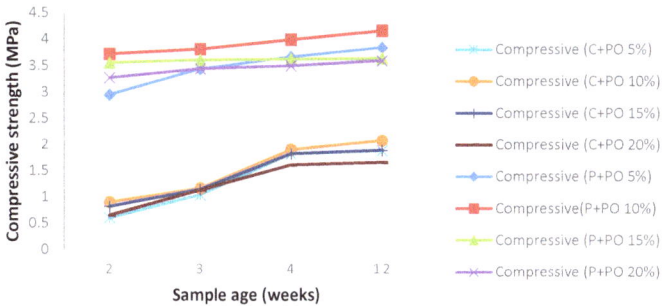

Fig. 1 Dry compressive strength of earthen blocks

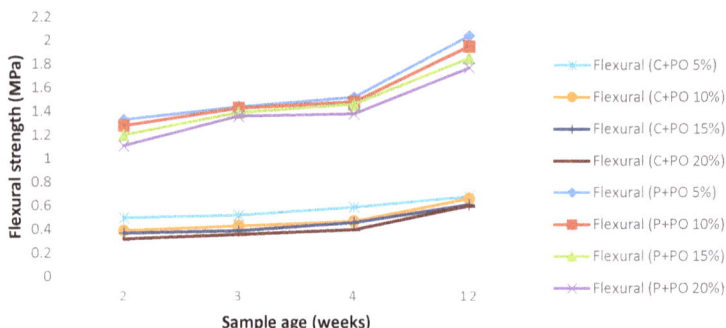

Fig. 2 Flexural strength of earthen blocks

Abbreviation:
C: Cement
P: Plaster
PO: Pozzolan

The results in Fig. 1 show that compressive strength increases slightly with increasing pozzolan dosage, for different curing times and for each hydraulic binder (cement or plaster), up to an optimum corresponding to 10% pozzolan, beyond which strength falls. By way of illustration and in terms of percentages, at an age of 28 days, increasing pozzolan content from 5% to 10% results in a rate of increase in compressive strength of the order of 8.87% and 3.84% respectively for plaster-based and cement-based blocks. However, an increase in pozzolan content from 10% to 20% results in a decrease in compressive strength of 12.41% for plaster blocks and 15.04% for cement blocks.

Analysis of Fig. 2 shows that the flexural strength of earth blocks decreases slightly with increasing pozzolan content, irrespective of the binder used or the curing time. For example, at an age of 28 days, increasing pozzolan content from 5% to 20% causes a reduction in flexural strength of around 10.14% for plaster blocks, and a reduction of around 32.2% for cement-based blocks

Fig. 1 and Fig. 2 also illustrate that plaster blocks have higher mechanical strength (compressive strength and flexural strength) than cement blocks. In other words, the use of plaster as a hydraulic binder enables the best mechanical strength to be obtained. For example, at an age of 28 days, the use of plaster induces a variation in compressive strength of 52.3%, 52.4%, 49.7%, and 52.4% for pozzolan contents of 5%, 10%, 15%, and 20%, respectively, compared with cement-stabilized blocks. Under the same conditions, flexural strength varies by 61.2%, 68.2%, 68.5%, and 71%, respectively.

The final point to be drawn from Fig. 1 and Fig. 2 concerns the effect of curing time on mechanical strength. This strength clearly increases irregularly with increasing curing time. Compressive strength increases rapidly between 14 days and 21 days, but more moderately between 21 days and 84 days. In addition, the evolution of mechanical strength is faster for cement-based blocks than for those made of plaster.

Hydric behavior. Table 6 illustrates the variation in the total water absorption of earth blocks as a function of pozzolan content, hydraulic binder type, and water immersion time, based on the different formulations proposed.

Emerging Research in Materials for Environment, and Civil Infrastructure - GeoME 5.5   Materials Research Forum LLC
Materials Research Proceedings 58 (2026) 32-39   https://doi.org/10.21741/9781644903933-5

Table 5 Total water absorption as a function of binder type, additive content and immersion time

[%] Pozzolan	Absorption in [%] (Plaster) Immersion time (Hours)				Absorption in [%] (cement)			
	24	48	72	96	24	48	72	96
5	7,27	10,86	12,74	14,85	7,59	11,34	15,78	20,27
10	8,10	12,65	13,86	15,85	8,25	13,56	17,33	20,80
15	9,34	13,1	14,95	16,94	8,65	14,08	17,82	21,07
20	10,19	13,52	15,06	17,2	9,75	15,09	18,73	21,32

The results in table 5 show that the rate of water absorption increases with increasing pozzolan dosage for different curing times and binder types.

We also note that plaster-based blocks are characterized by lower total absorption than cement-based blocks throughout all immersion periods and for all pozzolan contents. In addition, the water absorption behavior of blocks differs according to the type of binder used. Indeed, in plaster blocks, cracks begin to form from the 96th hour, with disintegration of their material at the surface in contact with water, which makes it impossible to measure their masses correctly, unlike cement blocks, which retain their shape (Fig. 3 and Fig. 4).

The final observation to be drawn from Table 5 concerns the effect of immersion time on the hydric behavior of the blocks. The total water absorption rate increases with increasing immersion time.

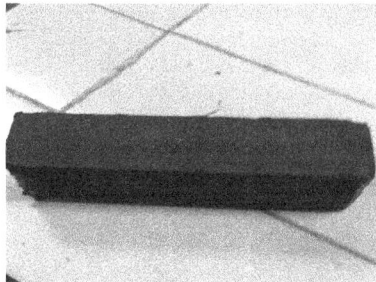

Fig. 3 Cement-based earth block after water absorption test

Fig. 4 Plaster-based earth block after water absorption test

Thermal behavior. Measuring the thermal conductivity of building materials enables us to select the most efficient materials for insulation. Insulating materials limit heat loss, thereby reducing heating and air-conditioning requirements, saving energy, and creating a comfortable environment.

Table 6 Thermal conductivity values of earth blocks for different binders and as a function of pozzolan content

[%] Pozzolan	thermal conductivity of plaster [$W. m^{-1}K^{-1}$]	thermal conductivity of cement [$W. m^{-1}K^{-1}$]
5	0,41	0,37
10	0,38	0,36
15	0,37	0,34
20	0,34	0,30

Table 6 shows the variation of the thermal conductivity of earth blocks for the different formulations studied. The incorporation of natural pozzolan significantly reduces thermal conductivity. For example, increasing the pozzolan content from 5% to 20% reduces thermal conductivity by about 17.1% in plaster blocks and 18.9% in cement blocks. In addition, cement blocks have a lower thermal conductivity than plaster blocks.

Discussion

The results of this experimental study show that increasing the pozzolan content causes a slight increase in compressive strength up to an optimum corresponding to 10% pozzolan, a decrease in flexural strength, a significant reduction in thermal conductivity and a drop in total water absorption. The natural pozzolans used in this study have a particle size of 5 mm or less, and are therefore highly porous. Their incorporation into the blocks therefore helps to increase porosity in the clay matrix. In general, the presence of pores increases the volume of gas in the sample, creating zones of weakness in the structure. On the one hand, this reduces heat transmission and consequently the thermal conductivity, and on the other, it makes the material brittle and sensitive, resulting in a reduction in mechanical and water resistance.

The results also confirm that curing time improves the mechanical strength of blocks. In other words, mechanical strength increases with increasing curing time. This is because, during curing, the various components of the blocks move closer together and become denser, forming bonds that strengthen the structure, and as a result, porosity is reduced, leading to an increase in mechanical strength.

This research also highlighted the impact of the type of hydraulic binder on block behavior. We observed that plaster significantly improved the mechanical and hydric performance of the blocks compared to those stabilized with cement. This can be explained by the fact that plaster adheres well to the constituent elements of the samples and existing pores in the blocks.

In the light of the results obtained throughout this experimental study, the following conclusions can be drawn:

- Plaster treatment improves mechanical strength compared with cement treatment.
- Mechanical strength increases with curing time. This increase is rapid for cement-based blocks.
- The addition of pozzolan leads to a reduction in thermal conductivity; this is due to the increased porosity in the clay matrix.
- For the different binders and for each age, compressive strength increases with increasing pozzolan content up to an optimum of 10%. Beyond this threshold, however,

strength begins to decrease slightly. On the other hand, flexural strength decreases with increasing pozzolan content.

- The total water absorption of cement- or gypsum-based blocks increases as pozzolan content rises. It also increases with increasing immersion time.
- Despite the lower absorption rate of plaster-based blocks compared to cement-based blocks, absorption still increases with immersion time.

References

[1] C. Arvind, K. Geetanjali, Renewable energy technologies for sustainable development of energy efficient building. Alexandria Engineering Journal 57(2), 655-69 (2018). https://doi.org/10.1016/j.aej.2017.02.027

[2] B. Mette, B. Jim, H. Chris, I. Lars Lonsmann, Time is running out for sand. Nature 571(7763), 29-31 (2019). https://doi.org/10.1038/d41586-019-02042-4

[3] A. Fernando, P. Esther, G. Rafael, G, Characterization of the mechanical and physical properties of stabilized rammed earth: A review. Construction and Building Materials 325(mars), 126693 (2022). https://doi.org/10.1016/j.conbuildmat.2022.126693

[4] N. Rouizem, The modernization of raw earth in Morocco: Past experiments and present. History of Construction Cultures, 1ere éd., 585-89 (2021). https://doi.org/10.1201/9781003173434-188

[5] Institut Marocaine de Normalisation (IMANOR) : Ouvrages en maçonnerie de petits éléments : Parois et murs. https://www.equipement.gov.ma/Ingenierie/Normalisation-et-Reglementation-technique/Documents/10.1.045.pdf (2020)

[6] A. Belabid, H. Akhzouz, H. Elminor, H., Characteristics of traditional building materials and techniques based on earth, stone and timbe: An overview and focus on Morocco. Journal of Engineering Research and Technology 11 (2023)

[7] R. Karka Bozabe, D. Taipabe, Modèle de construction d'habitat en terre : Cas d'adobe manuel en Afrique au sud du Sahara [Land habitat construction model : Adobe manuel case in Africa, South of the Sahara]. International Journal of Innovation and Applied Studies 26(4), 883-87 (2019)

[8] GC. Bailly, Y. El Mendili, A. Konin, E. Khoury, Advancing earth-based construction: A comprehensive review of stabilization and reinforcement techniques for adobe and compressed earth blocks. Eng 5(2), 750-83 (2024). https://doi.org/10.3390/eng5020041

[9] A. Bouyahayaoui, M. Cherkaoui, L. Abidi, T. Cherradi, Mechanical and chemical characterization of pozzolan of middele Atlas in Morocco. International Journal of GEOMATE 14 (41), 126-134 (2018). https://doi.org/10.21660/2018.41.91013

Emerging Research in Materials for Environment, and Civil Infrastructure - GeoME 5.5   Materials Research Forum LLC
Materials Research Proceedings 58 (2026) 40-46   https://doi.org/10.21741/9781644903933-6

Seismic evaluation of compressed earth blocks using acoustic waves: Case of Chichaoua Taroudant

Abderrahmane Jouhar[1,a] *, Mohammed Cherraj[1,b], Mokhfi Takarli[2,c], Fatima Allou[2,d], Driss El Hachmi[1,e]

[1]MSME, Faculty of science Mohammed V University in Rabat Morocco

[2]University of Limoges, GC2D, Egletons, France

[a]Abderrahmane.jouhar@gmail.com, [b]m.cherraj@um5r.ac.ma, [c]mokhfi.takarli@unilim.fr, [d]fatima.allou@unilim.fr, [e]d.elhachmi@um5r.ac.ma

Keywords: Compressed Earth Block, Ultrasound, Vulnerability, Seismic

Abstract. Compressed earth blocks (CEB), also referred to as BTC, are long-used earthen materials valued for durability and energy efficiency. On 8 September, Morocco was struck by a devastating earthquake that claimed nearly 3,000 lives and severely affected the provinces of Al-Haouz, Chichaoua, and Taroudant, leading to the suspension of educational activities in roughly 40 municipalities. To better understand how local materials behave under seismic demand, raw soils were sampled in the village of Amskrdad within the impacted Chichaoua-Taroudant region and used to fabricate an unstabilized compressed earth brick representative of local practice. The mechanical response was then probed by inducing ultrasonic waves along the x- and y-axes to measure pulse velocity and time of flight (TOF), from which the signal-to-noise ratio (SNR) was computed for each record. Tests carried out on a Pundit PL200 (Proceq) in fully non-destructive mode yielded an elasticity modulus of 2819.92 MPa and a Poisson's ratio of 0.27. They enhance our understanding of stiffness and lateral strain behavior of CEB under seismic loading and offer practical data on the structural reliability of traditional earth construction in seismically active regions.

Introduction

Building and have been appreciated for their strength, mass, and relatively benign environmental impacts [1]. In Morocco, they have been used extensively in the area of housing due to the enhanced environmental envelope that they offer while having low environmental impacts [2,3].

On 8 September 2023, a Mw 6.8 earthquake in the High Atlas exposed how such earthen systems can perform in seismically active settings, with nearly 3,000 fatalities and heavy damage reported across Al-Haouz and Chichaoua-Taroudant [4].

In addition to the immediate damage, it led to a focus on the seismic integrity of CEB buildings in the area as well as the need for site-specific information on the mechanical properties of unstabilized blocks formulated from local soils.

This paper answers these needs. The raw soils were collected from a rural village of Amskrdad (Chichaoua-Taroudant) as sources. From these soils, unstabilized CEB were made and then ultrasonic testing carried out. The question being asked here remains very clear: can conventionally made unstabilized CEB possess adequate strength and stiffness to withstand seismic loads? If the answer turns out negative, then what needs to be investigated next is the modification that fits improved granulometry/moisture management, reduced stabilizing components, and details. Our contribution is twofold: (1) a reproducible, materials-based assessment of the seismic adequacy of locally produced, unstabilized CEB, and (2) evidence-grounded recommendations for reconstruction that preserve the environmental and cultural value of earth construction while improving safety and resilience in the affected areas.

Experimental details

The study followed a series of essential steps to analyze the mechanical properties of CEB obtained from the region affected by the September 8, 2023, earthquake. Soil specimens were collected in the vicinity of the village of Amskrdad, situated in the Chichaoua-Taroudant province, one of the most severely impacted areas. These samples were collected from locations traditionally used by local residents as a source of earth for construction.

After collection, the raw soil was left to air-dry to eliminate residual moisture before being processed into CEB (Fig. 1).

***Fig. 1.** Satellite image of the area where the soil samples were taken in the village of Amskrdad, Chichaoua Tar-oudant, Morocco.*

The compacting of the soil sample was done by a hydraulic press that had a compression force of 300 kN. The force created standard bricks that were $17 \times 12 \times 8$ cm.

The CEB were finally dried uniformly by heating them in an oven at 105°C for 72 hours [5].

Subsequently, the CEB underwent non-destructive testing using the Pundit PL-200 ultrasonic device that had high accuracy [6].

The device measures pulse velocity and the time of flight in both longitudinal and transverse directions (X and Y), respectively.

The ultrasonic waves were produced for the two samples. Analysis of the waveform data made it possible to determine the signal-to-noise ratio for each of the measurements.

The collected data were used to compute two key mechanical properties: the elastic modulus and Poisson's ratio, utilizing the PL-LINK software [7].

Since these tests were conducted without causing harm to the material, these tests give an important indication of the strength of the CEB when faced with seismic actions.

The last analysis of the tests carried out is to determine whether these CEB blocks need enhancements for them to be fit for use in earthquake zones (Fig. 2).

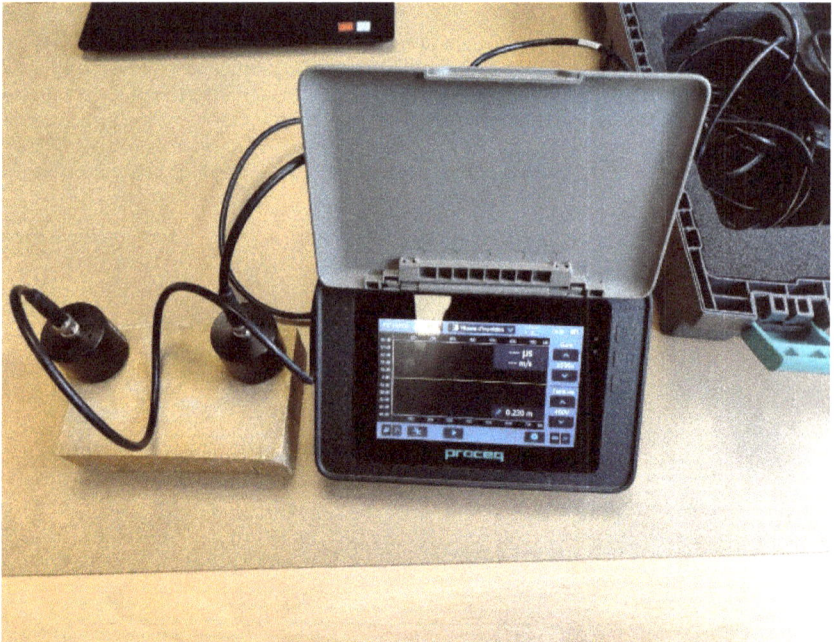

Fig. 2. Pundit PL-200 and P- and S-wave transducers

Results and discussion

Gain and voltage combination measurements were recorded in the x and y directions, with transducer positioning suited to enable wave transmission [8].

Time of flight (TOF) and pulse velocity were recorded using the PL-Link software, the data saved as Excel-readable files for later analysis [9].

The presence of higher frequencies during testing depicted enhanced material absorption of ultrasonic waves and lower frequencies showed deeper wave penetration [10].

Voltage and gain increase of the ultrasonic transmitter in CEB, as observed in Figure 3, resulted in power amplification of the signal and thus increased the speed of wave propagation as well as decreased TOF.

The relationship among gain, voltage, TOF, and SNR confirms that at higher voltages (300 - 400 V), the SNR significantly increased, delivering better wave transmission with less noisy interference.

Of special note at 300 V, the SNR was more than 50 dB, confirming the accuracy of measurements in TOF. Moreover, as gain increased, TOF exhibited a negligible decrease due to enhanced signal power, which allowed for faster wave transmission through the material.

Comparison of x and y directions revealed anisotropic wave propagation, with x-direction possessing greater TOF and lesser velocity compared to the y-direction.

This is a sign of small material inhomogeneities which may affect wave transmission. Figure 4 indicates that using higher voltage or gain resulted in greater TOF, and velocity and SNR also increased.

Emerging Research in Materials for Environment, and Civil Infrastructure - GeoME 5.5   Materials Research Forum LLC
Materials Research Proceedings 58 (2026) 40-46   https://doi.org/10.21741/9781644903933-6

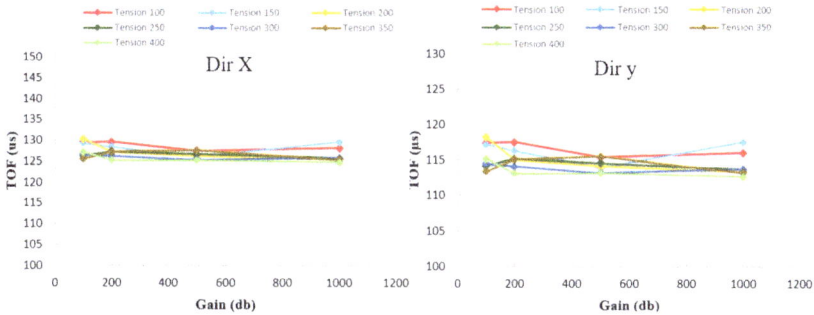

Fig.3. TOF as a function of gain and voltage

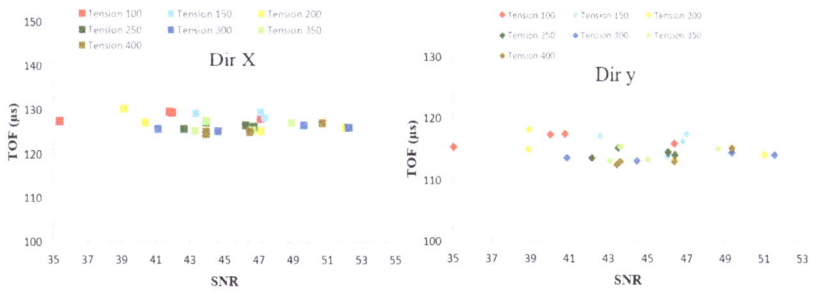

Fig.4. TOF as a function of SNR for transducer 54

The higher the SNR value indicated that the noise levels were lower, hence better clarity of the signal. Choosing an optimum value of TOF was important in estimating the Young's modulus of elasticity (E) and the value of Poisson's ratio (υ), as both estimates were affected by having higher values of SNR. As per Figure 5, the values of 126.65 μs for x-direction and 114.78 μs for y-direction were used as TOFs where measurements were stopped due to stabilization [11].

Based on the velocities V(X) and V(Y) given in Table 1, Young's Modulus formula (E = 2819.92 MPa) and the value of Poisson's Ratio formula (υ = 0.27) were derived from the formula involving the velocities of P-waves and S-waves given by [12]:

$$VP = \sqrt{\frac{E(1-\nu)}{\rho(1+\nu)(1-2\nu)}} \tag{1}$$

$$VS = \sqrt{\frac{E}{2\rho(1+\nu)}} \tag{2}$$

With:

VL: The propagation velocity of the P-wave
VT: The propagation velocity of the S-wave
E: Young's Modulus (Pa)
σ: Poisson's Ratio
ρ: Density (kg/m³)

The Young's modulus and Poisson's ratio values can be inferred from these equations as follows:

$$\nu = \frac{VP^2 - 2VS^2}{2(VP^2 - VS^2)} \tag{3}$$

$$E = 2\rho VS^2(1 + \nu) \tag{4}$$

Table 1. *Calculation of velocities V (0) and V (90).*

	TOF (µs)	Distance [m]	velocities
V(0)	126,65	0,17	1342,20
V(90)	114,78	0,12	1045,40

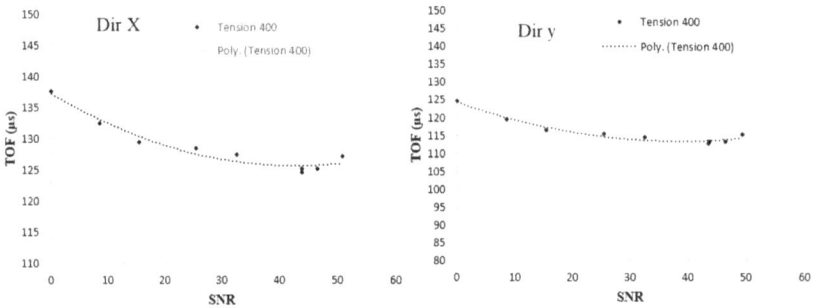

Fig.5. *TOF as a function of SNR for transducer 54*

These findings show that CEB have mechanical properties that depend on SNR and pulse velocity. Therefore, CEB may have a moderate degree of stiffness and elasticity. However, it may have inadequate structural strength as regards seismic activity in earthquake-prone areas [13].

Increased applied voltage and gain increased the speed of propagation of the waves and shortened the value of TOF, hence reduced attenuation of the measurements [14].

The higher values of SNR at higher gain settings confirmed the increased transmission of waves and hence proved the correctness of the mechanical values derived for seismic analysis [15].

The velocity discrepancies in V(x) and V(y) illustrate the material anisotropy that considers higher velocities in the x-direction than those in the y-direction.

This can be attributed to the composition of raw earth. Although the value of the Young's moduli range is quite moderate, it's still much lower than that of the stabilized forms. This indicates that CEB cannot have adequate stiffness to withstand the stress caused by an earthquake.

Additionally, the Poisson's ratio of 0.27 suggests moderate material compressibility, which could lead to substantial deformation under seismic forces if reinforcements are not introduced [15].

In addition to compacting force and water content, studies may investigate the use of stabilization methods such as cementing or fibers to improve CEB resilience against earthquakes.

Additional simulations based on earthquake conditions can also be carried out to confirm the validity of data collected from ultrasonic testing [16].

The above research findings highlight the importance of material enhancement as a factor that contributes to earthquake-safe construction processes in earthquake-prone regions.

Conclusions

The research work investigated the seismic resilience of CEB by ultrasonic tests to establish the significance of such tests in seismic loads. The Young's modulus and Poisson's ratio of unstabilized CEB from the earthquake-sensitive area of Chichaoua-Taroudant were indicated to be approximately in the range of moderate stiffness and elasticity. From the data on the speed of propagation of waves, anisotropic properties were observed, which reflect internal inhomogeneity of composition and overall density. In addition, it has been observed that higher transmitter voltage and gain settings increased penetration depth and quality of the signal received, reduced value of TOF, and increased levels of confidence in the derived values. However, these values of moduli of elasticity were lower than those usually given for the stabilized ones. Hence, it may be logically assumed that unstabilized CEB may not offer adequate earthquake loads.

The value of the derived Poisson's ratio of CEB unstabilized (= 0.27) offers an indication of probable compressibility and hence earthquake vulnerability. To offer improved seismic resilience, it may be imperative that subsequent research work on similar CEB incorporates light stabilization levels (minimum contents of cement and natural binders) and better control of compacting energy and water content. These tests may be followed by further academic simulations of earthquake parameters to validate the findings by suggesting improvements. As long as CEB remain robust and economic resources in seismic regions, it may be imperative that certain improvements as suggested above come into play. In addition, the research work raises localized mechanical properties of CEB unstabilized from being academic principles to being implementable actions useful for improved seismic reconstruction within earthquake-prone zones.

Acknowledgments

The authors wish to acknowledge with thanks the GC2D Laboratory, University of Limoges, Egletons, France, for providing the facilities to carry out this work. Furthermore, they gratefully acknowledge the financial support provided by the CNRST under the "PhD-Associate Scholarship PASS" program.

References

[1] Hema, C., Messan, A., Lawane, A., Soro, D., Nshimiyimana, P., & Van Moeseke, G. (2021). Improving the thermal comfort in hot region through the design of walls made of compressed earth blocks: An experimental investigation. Journal of Building Engineering, 38, 102148. https://doi.org/10.1016/j.jobe.2021.102148

[2] Jouhar, A., Hachmi, D. E., Moussaoui, R., Cherraj, M., Bergui, S. E, Zoubair, A.B. & Nchiti, E.M. (2024). Impact of stabilizers on the properties and energy efficiency of compressed earth blocks. Sustainable Mediterranean Construction (19), 93-97.

[3] Teixeira, E. R., Machado, G., P. Junior, A. D., Guarnier, C., Fernandes, J., Silva, S. M., & Mateus, R. (2020). Mechanical and thermal performance characterisation of compressed earth blocks. Energies, 13(11), 2978. https://doi.org/10.3390/en13112978

[4] Haddad, E. A., El Aynaoui, K., Ali, A. A., Arbouch, M., Saoudi, H., & de Araújo, I. F. (2024). Assessing the Economic Impacts of Al-Haou Earthquake: Damages and Recovery Strategy (No. 1981). Policy Center for the New South.

[5] Lourenço, P., de Brito, J., & Branco, F. (2002, September). The use of compacted earth blocks (BTC) in modern construction. In XXX IAHS World Congress on Housing: Housing Construction-an Interdisciplinary Task. September 9-13, Coimbra. Portugal.

[6] Fitri, F. A. (2018). Analisis Modulus Elastisitas Beton Dengan Menggunakan Alat Pundit Pl-200 (Doctoral dissertation, Universitas Brawijaya).

[7] Chaix, JF, Garnier, V., & Corneloup, G. (2006). Propagation d'ondes ultrasonores en milieux solides hétérogènes : analyse théorique et validation expérimentale. Ultrasons , 44 (2), 200-210. https://doi.org/10.1016/j.ultras.2005.11.002

[8] Tzelepi, N. (2014). Sample Size Effects on Ultrasonic Measurements of Elastic Moduli-Experimental and Theoretical Investigations. In Graphite Testing for Nuclear Applications: The Significance of Test Specimen Volume and Geometry and the Statistical Significance of Test Specimen Population. ASTM International. https://doi.org/10.1520/STP1578-EB

[9] Angrisani, L,& Moriello, R. S. L. (2006). Estimating ultrasonic time-of-flight through quadrature demodulation.IEEE transactions on instrumentation and measurement, 55(1), 54-62. https://doi.org/10.1109/TIM.2005.861251

[10] Suñol, F., Ochoa, D. A., & Garcia, J. E. (2018). High-precision time-of-flight determination algorithm for ultrasonic flow measurement. IEEE Transactions on Instrumentation and Measurement, 68(8), 2724-2732. https://doi.org/10.1109/TIM.2018.2869263

[11] Franco Guzmán, E. E., Meza, J. M., & Buiochi, F. (2011). Measurement of elastic properties of materials by the ultrasonic through-transmission technique.

[12] Benaboud, S. (2022). Evaluation du vieillissement et de l'endommagement des matériaux bitumineux par modélisation hétérogène et mesures acoustiques (Doctoral dissertation, Université de Limoges).

[13] Labiad, Y., Meddah, A., & Beddar, M. (2022). Physical and mechanical behavior of cement-stabilized compressed earth blocks reinforced by sisal fibers. Materials Today: Proceedings, 53, 139-143. https://doi.org/10.1016/j.matpr.2021.12.446

[14] Bernat-Maso, E., Teneva, E., Escrig, C., & Gil, L. (2017). Ultrasound transmission method to assess raw earthen materials. Construction and Building Materials, 156, 555-564. https://doi.org/10.1016/j.conbuildmat.2017.09.012

[15] Ekin, N. (2025). The relationships between ultrasonic P and S wave velocities and resistivity in reinforced concrete. Construction and Building Materials, 479, 141475. https://doi.org/10.1016/j.conbuildmat.2025.141475

[16] Santos, P., Júlio, E. N. B. S., & Santos, J. (2010). Towards the development of an in situ non-destructive method to control the quality of concrete-to-concrete interfaces. Engineering Structures, 32(1), 207-217. https://doi.org/10.1016/j.engstruct.2009.09.007

Emerging Research in Materials for Environment, and Civil Infrastructure - GeoME 5.5   Materials Research Forum LLC
Materials Research Proceedings 58 (2026) 47-53   https://doi.org/10.21741/9781644903933-7

Comparative analysis of the stability performance of solid and perforated brick masonry structures

Chaimae Khanfri[1,a] *, Ouadia Mouhat[2,b], Younes El Rhaffari[2,c],
Fatima El Mennaouy[1,d]

[1]Civil Engineering and Environment Laboratory (LGCE), Mohammadia Engineering School,
Mohammed V University, Rabat, Morocco

[2]Civil Engineering and Environment Laboratory (LGCE), School of Technology-Sale,
Mohammed V University, Rabat, Morocco

[a]chaimae.khanfri93@gmail.com, [b]ouadie.mouhat@est.um5.ac.ma, [c]ayounes1@hotmail.com,
[d]fatimaelmennaouiy@gmail.com

Keywords: Masonry Structure, Stability Performance, Repeated Load Cycles, Solid Panel, Abaqus

Abstract. Around the world, unreinforced masonry (URM) walls remain widely used, particularly as infill walls, often featuring openings for the installation of doors and windows. These structures are particularly vulnerable to lateral loads, highlighting the need to analyze their structural behavior to propose suitable reinforcement solutions. In this article, the stability performance of masonry structures was analysed under repeated load cycles through simulations performed in Abaqus, studying their energy dissipation capacity and sensitivity to seismic demands. Simplified microscopic simulations are performed founded on the finite element technique, combining the Concrete Damage Plasticity (CDP) theory for brick modeling and a cohesive approach to analyze the failure mechanism of the joints. The study focuses on two masonry wall configurations under cyclic loading: a solid panel (without openings) and a panel with an opening. The objective is to compare their response by analyzing the deformation modes as well as the corresponding load–displacement curves. The findings show that the presence of openings reduces the overall wall strength and generates localized cracking around these openings, which constitute weak zones.

Introduction

The use of local materials has currently become a valued objective in the field of construction in the context of sustainable development. These materials, are natural their use does not have any negative impact on the environment [1]. Among the resources most used for years in the construction sector, the masonry which consists of stones and mortar [2], it has several advantages, it is an ecological and durable material and also provides high comfort. Moreover the masonry is characterized by a high resistance to heat and a strong acoustic insulation [3]. Masonry, for centuries has been an important component of the architectural heritage of Morocco, is the main focus of this study. Masonry structures possess excellent strength against vertical compression but are very sensitive to in-plane lateral loading, including severe wind gusts and earthquake [3]. During seismic events, this type of structure is subjected to widespread damage [4]. In this paper, a three-dimensional numerical model was employed with the finite element method. The modeling technique used in this study is simplified micro-modeling, with the aim of investigating the stability performance of brick structures, both apertured and non-apertured, under periodic loads. A 3D model for the wall was created using the ABAQUS software. The aim is to examine the influence of apertures on the rupture mechanism of the structure. Mode of deformations and load-displacement curves are presented and investigated for every case. The findings shown lead to the conclusion that the introduction of an opening lowers the total capacity of the wall and also affects its failure mode.

Modeling methodology

A three-dimensional analysis is performed through a simplified microscopic simulation founded on the finite element method. This process provides precise outcomes. while minimizing computational time. The numerical modellings are executed with the Abaqus software. The bricks are modeled with dimensions up to half of the connections in order to simulate the bricks' and mortar's interaction, characterized by a modified elastic modulus accounting for their structural and dimensional properties. The Concrete Damage Plasticity (CDP) model, used to introduce the global complex mechanical response of masonry. A surface-oriented cohesive approach is employed for the simulation of mortar, it is modeled using an interface element with no thickness. This modeling strategy, founded on the combination of various constitutive models, allows for: Improved representation of the discontinuity of masonry, more realistic reproduction of failure mechanisms, and accurate monitoring of the development of deformation throughout the different stages of loading.

Constitutive models

Combined Plasticity and Damage Model (CDP). The Concrete Damage Plasticity (CDP)model is a continuous model incorporating two major deformation processes: cracking due to compression failure and tension [9]. Initially developed to characterize the nonlinear response of brittle-like materials such as concrete subjected to cyclic and dynamic stresses [10], this model has also proven effective when applied to masonry [11]. The parameters of this model were determined based on previous research [7]. The dilation angle was set to 10°, while the eccentricity value was defined as 0.1. The ratio between the biaxial and uniaxial compressive strengths (Fbo/Fco) was taken as 1.16. Moreover, a small viscosity coefficient of 0.002 was adopted to optimize the convergence of the numerical simulations.

Cohesive approach. The cohesive method is a modeling technique used in FE-based simulation to analyze the behavior of interfaces. The mortar is considered as a zero-thickness interface, and a traction-separation law defines the cohesive model. This means that a relationship is established between the applied pressures (normal and shear) as well as the corresponding movements between the two surfaces (or blocks). The contact type used for this study is "surface-to-surface", and then defining the normal and tangential behaviors (including friction, typically based on a Coulomb friction model). Rfer to Table 1.

Numerical study

Materiel Properties. The structural characteristics of the materials that make up the masonry, are presented in Table 1. To ensure rigid boundary conditions, the wall is fixed at the top and bottom using rigid steel beams with high modulus of elasticity, thereby ensuring uniform stress distribution and adequate stability during the simulation.

Emerging Research in Materials for Environment, and Civil Infrastructure - GeoME 5.5   Materials Research Forum LLC
Materials Research Proceedings 58 (2026) 47-53   https://doi.org/10.21741/9781644903933-7

Table 1. The properties of materials

Property	Corresponding Value
- Elastic property of the brick units Eu (MPa)	2400
- Brick unit density (g/mm3)	2.4E-09
- Poisson's ratio of the bricks μ	0.15
- Elastic property of the mortar Em (MPa)	780
-Extended blocks Eadj (MPa)	5884
Interfacial characteristics:	
- Normal-direction nominal stress Knn (N/mm3)	82
- Shear-orientation I nominal stress Kss (N/mm3)	36
- Shear-orientation II nominal stress Ktt (N/mm3)	36
- Friction factor	0.7

Equation (1) [2] is used to calculate the expanded bricks' effective elastic modulus, taking into account the construction's dimensions and the initial elastic moduli of the bricks and mortar.

$$E_{adj} = \frac{HE_bE_m}{nh_bE_m + (n-1)h_mE_b} \tag{1}$$

Validation of the numerical model. Model validation is a crucial step in assessing the reliability of the numerical approach adopted for analyzing the dynamic reaction of brick constructions. In this context, a three-dimensional model is developed employing the ABAQUS program and the FE approach (detailed in section 2). The analyzed wall has the same dimensions as that in reference [5], with a proportion of length to height of about 0.99 (990 mm in length and 1000 mm in height). It consists of 18 rows of bricks. Two rigid beams are placed at the ends of the wall. to provide anchorage and accurately replicate the boundary conditions from reference [5]. A force in the vertical direction of 2.12 MPa is also exerted via the top beam. fig.1-a. The structural characteristics of the bricks, mortar, as well as the interface elements, are displayed in the Table 1. The modified elastic modulus of the extended blocks is calculated using Equation (1). Fig.1-b shows the numerical data obtained from the modeling, which demonstrate a significant conformity with the experimental data reported in reference [2], thus confirming the reliability of the employed numerical strategy.

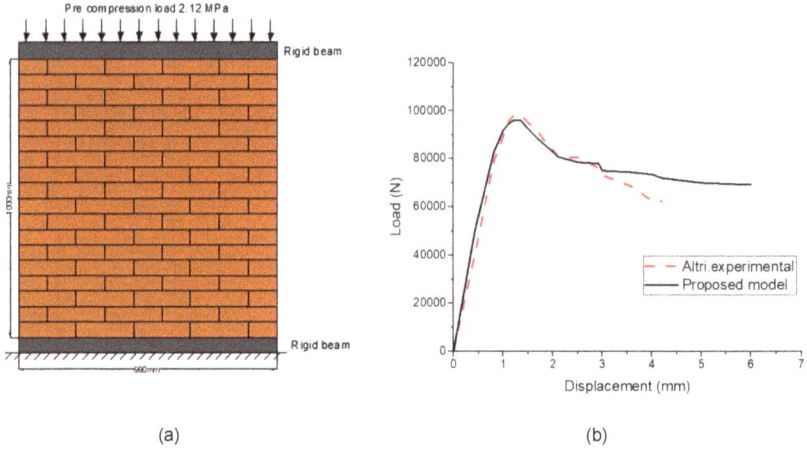

(a)   (b)

Fig. 1 Validation of numerical model: (a) configuration and loading conditions, (b) The recommended model's load distribution curve and the experimental outcomes [5], under a vertical pressure of 2.12 MPa.

Evaluation of the masonry wall subjected to dynamic load. The study of the wall's response under dynamic loading was carried out using the experimental results published in reference [6], in order to identify the failure mode and examine the way that apertures impact the dynamic behavior of the wall. For this purpose, the wall described in this reference was modeled in Abaqus software, employing the previously outlined modeling methodology. This simulation uses simple microscopic modeling and is founded on a numerical FE model, as mentioned earlier. The same procedure is adopted here under cyclic load-ing, integrating the stones' structural features as well as those of the inter-faces to simulate more precisely the performance of masonry elements and joints facing this kind of load. The studied wall is built according to the classic cross-joint masonry method shown in Fig. 2. It measures 1206 mm in length and 800 mm in height. Steel beams support the two upper and lower extremities of the wall. The top beam imposes an initial vertical pressure of 60 kN. Thereafter, this beam is loaded cyclically, which causes lateral motions but limits rotation. Throughout the simulation, the vertical pressure remains constant. The bottom beam is fastened to the earth, thus ensuring loading conditions identical to those described in the reference [6]. Two numerical simulations were performed: the first for a wall without openings, and the second with an opening created in the wall's middle, representing 5% of the total wall surface area. The observations from both numerical analyses were evaluated with the test outcomes. Fig. 2 and Fig.3 show the different stages of the simulation of the wall without and with the opening.

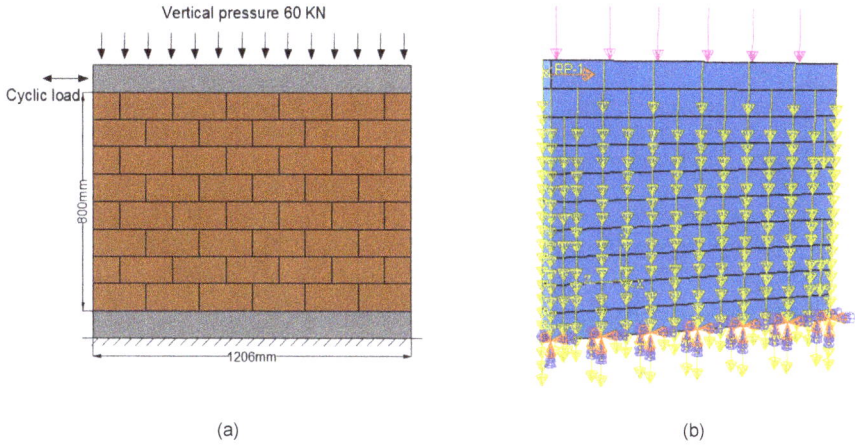

(a)   (b)

Fig. 2 Wall without opening under cyclic load; (a) wall configuration and loading conditions, (b) boundary conditions in Abaqus.

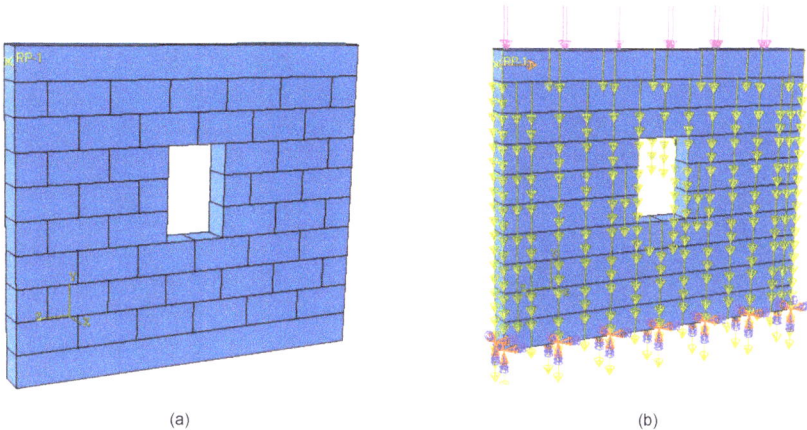

(a)   (b)

Fig. 3 Wall with opening under cyclic load; (a) wall configuration and loading conditions, (b) boundary conditions in Abaqus

Results

After the numerical analysis, the findings indicate that the deformation process of the brick wall exposed to a periodic load is distinguished by the formation of fractures with a diagonal orientation. This form of deformation results from the successive pressures of shear and stretching movements produced by the imposed periodic load. For the wall without openings, the initial cracks form at the edge of the wall and quickly spread toward the middle. These cracks draw two diagonal axes that intersect, forming an X-shaped motif. This observation is consistent with the experimental results reported in reference [6] (see Fig4-a, and Fig.4-b). This is proof of how

Emerging Research in Materials for Environment, and Civil Infrastructure - GeoME 5.5   Materials Research Forum LLC
Materials Research Proceedings 58 (2026) 47-53   https://doi.org/10.21741/9781644903933-7

masonry structures are susceptible to horizontal stresses. The fractures tend to develop in the joints' orientation of the mortar, and as they develop, the condition of the wall continues to deteriorate and can lead to additional damage without support. For the opening wall, the cracks are more severe. The same pattern of diagonal failure is observed; however, the center opening is a zone of weakness where the cracks extend quickly (Fig4-c). Fig 5 shows the load-displacement curves for both simulations along with the experimental curve from reference [6]. It is evident that the opening significantly affects the wall's capacity. The solid wall has a highest cyclic load resistance of 35 kN, while the opening-containing wall can only sustain 22 kN.

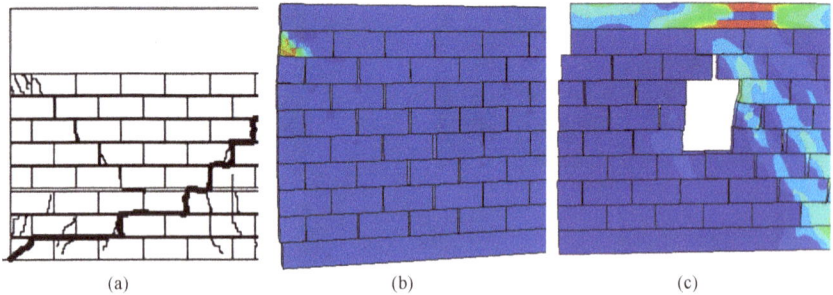

(a)   (b)   (c)

Fig. 4 Comparing test results [6] and current numerical observations in cyclic loading: (a) failure mechanism [6], (b) current numerical results of wall without opening, (c) existing numerical outcomes of wall with opening

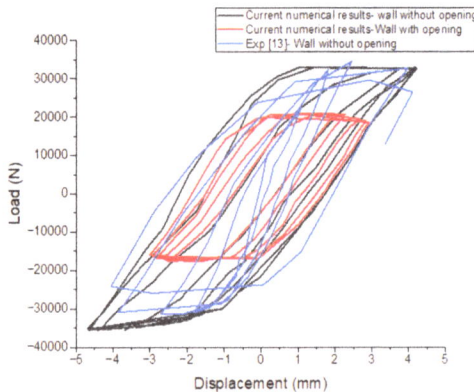

Fig. 5 Comparison of force-displacement curve under cyclic load.

Conclusion

This paper is a study on the dynamic response of cyclically loaded brick walls with and without openings and is divided into two quite dissimilar sections. The first section is concerned with a validation study for checking the effectiveness of the technique used, and the second section discusses the impact of apertures on wall behavior. Analysis such as this helps to advance the knowledge of such buildings, allowing for the implementation of appropriate interventions to strengthen them and in particular minimize their seismic vulnerability. The results significantly stress that:

• Cyclic loading is most susceptible in stone masonry walls, which can only support light loads and exhibit extremely large cracks.

• Openings in masonry walls significantly influence the failure mechanism of masonry walls, making them less rigid. Diagonal fractures, that increase significantly with openings, appear to be the most frequent way of breaking. Holes show the point where the structure is vulnerable, which speeds up the transmission of deformations.

References

[1] R. El Nabouch1, Q.-B.Bui1, P.Perrotin1, O.Plé1, J.-P.Plassiard1. Numerical Modeling of Rammed Earth Structures: Analyses and Recommendations.2015

[2] Kurdo F. Abdulla a, Lee S. Cunningham a, Martin Gillie b. Simulating masonry wall behaviour using a simplified micro-model Approach. (2017). https://doi.org/10.1016/j.engstruct.2017.08.021

[3] Meillita Finite element modelling of unreinforced masonry (URM) wall with openings: studies in Australia. The Proceedings of 2nd Annual International Conference Syiah Kuala University (2012).

[4] Danna Darmayadi, Muhamad Rusli. Element modeling of masonry wall with opening under lateral force (2018). https://doi.org/10.30659/jacee.1.2.71-87

[5] Antonio Maria D'Altri a, Stefano de Miranda a, Giovanni Castellazzi a, Vasilis Sarhosis b. A 3D detailed micro-model for the in-plane and out-of-plane numerical analysis of masonry panels (2018) https://doi.org/10.1016/j.compstruc.2018.06.007

[6] Vladimir G. Haach, Graça Vasconcelos, and Paulo B. Lourenço. Experimental Analysis of Reinforced Concrete Block Masonry Walls Subjected to In-Plane Cyclic Loading. Journal of Structural Engineering (APRIL 2010).

[7] Suraj D. Bhosale, Atul K. Desai. Simulation of Masonry Wall using Concrete Damage Plasticity Model.2019

[8] Elide Nastri , Michela Tenore , Paolo Todisco. Calibration of concrete damaged plasticity materials parameters for tuff masonry types of the Campania area. 2023 https://doi.org/10.1016/j.engstruct.2023.115927

[9] Noor-E-Khuda S. An Explicit Finite-Element Modeling Method for Masonry Walls Using Continuum Shell Element.

[10] Lubliner, J.; Oliver, J.; Oller, S.; Onate, E. "A Plastic-Damage Model". Int. J. Solids Struct. 1989, 25, 299-326. https://doi.org/10.1016/0020-7683(89)90050-4

[11] Valente, M.; Milani, G. "Damage Assessment and Collapse Investigation of Three Historical Masonry Palaces under Seismic Actions". Eng. Fail. Anal. 2019, 98, 10-37. https://doi.org/10.1016/j.engfailanal.2019.01.066

Emerging Research in Materials for Environment, and Civil Infrastructure - GeoME 5.5   Materials Research Forum LLC
Materials Research Proceedings 58 (2026) 54-61   https://doi.org/10.21741/9781644903933-8

Thermal performance of plaster composites reinforced with date palm fibers: Experimental and numerical analysis

Youssef Khrissi[1,a] *, Mohamed Char[1,b], Amine Tilioua[1,c] and Mohamed Azrour[2,d]

[1]Research Team in Thermal and Applied Thermodynamics (2.T.A.). Department of Engineering Sciences, Faculty of Sciences and Techniques Errachidia, Moulay Ismaïl University of Meknès, B.P. 509, Boutalamine, Errachidia, Morocco

[2]Laboratory of Materials Engineering for the Environment and Natural Resources, Faculty of Science and Technology, Moulay Ismail University, Errachidia, 52000, Morocco

[a]yo.khrissi@edu.umi.ac.ma, [b]m.char@edu.umi.ac.ma, [c]a.tilioua@umi.ac.ma, [d]m.azrour@fste.umi.ac.ma

Keywords: Sustainable Materials, Gypsum, Date Palm Fibers, Thermal Insulation, Numerical Simulation

Abstract. Trends towards sustainable construction growing interest in sustainable design and development has inspired a new generation of innovative composites for energy-efficient buildings. Date palm fibers are an abundant and widely available agricultural waste in arid and desert regions. Several works have shown the advantages of plant fibers for thermal insulation, but have not yet notably reported on plasters with these plant fibers. In this study, the thermal performance evaluation of date palm leaf fiber and treated composites was studied towards their potential applications for building materials. The formulation was employed using the gypsum with levels of fiber (0% to 5%, mass of fiber) added. The thermal characteristics of these formulations were further investigated through experiments. These values were then included in a COMSOL Multiphysics numerical model related to heat transfer across a wall containing referred materials. Results showed that a clear decrease in heat conductivity occurred when the fiber content increased. 5% fiber addition had the best insulating performance, folded with the optimum. This enhancement is attributed to the inherent thermal insulating characteristics of the fiber components of the plant. The numerical analysis supported the results obtained experimentally by showing a notable decrease in heat transfer through the composite wall.

Introduction

The building industry in Morocco ranks among the top three energy-consuming sectors. This high demand poses a significant challenge to sustainable development policies and the shift towards a more energy-efficient model. Although some initiatives have been undertaken to enhance energy performance in this area, several constraints still hinder this evolution. These include the limited insulation quality of existing buildings, the shortage of suitable local resources, the absence of reliable databases on their properties, and a still limited understanding of the influence of hygrothermal characteristics of materials on indoor comfort and energy consumption. Therefore, educating building professionals about sustainable alternatives is becoming more and more important. Bio-based materials are becoming more and more popular in this regard. They have several applications in both new building and remodeling. Because of their exceptional capacity to control heat transfer and moisture, they stand out for their the capacity to efficiently keep the temperature inside balance and control moisture conditions [1]. As part of global commitments to reduce greenhouse gas emissions, public authorities also support the enthusiasm for these materials, setting targets of between 80% and 95% reduction by 2050 and a 55% reduction by 2030 compared to 1990 levels. An inventive strategy in this regard is the substitution of some of the traditional materials with additives derived from biomass. These remedies promote affordable,

Emerging Research in Materials for Environment, and Civil Infrastructure - GeoME 5.5   Materials Research Forum LLC
Materials Research Proceedings 58 (2026) 54-61   https://doi.org/10.21741/9781644903933-8

sustainable housing while reducing their adverse environmental effects. Extensive research has been conducted on recovering agricultural residues and natural by-products to improve the thermal, mechanical, acoustic, and hydraulic behavior of construction materials while consuming fewer energy-intensive resources. These efforts have led to the development of a new generation of high-performance, environmentally friendly composites. Traditional materials like plaster, gypsum, and clay are becoming more and more popular because of their minimal impact on the environment. These materials are now stronger, more resilient, and more insulating thanks to plant fibers. The thermophysical properties of renewable fiber composites make them suitable for modern sustainable building. To make gypsum, natural gypsum is ground and calcined. It is widely used to create prefabricated parts such as panels and tiles. Its performance can be improved by adding natural reinforcements, as many studies have investigated. Khanfer et al. [2], for instance, demonstrated that cellulose fibers enhance gypsum's mechanical resistance by delaying the onset and propagation of cracks. The impact of granular or fibrous natural insulating additives on the mechanical and thermal properties of gypsum-based composites was investigated by Maaloufa et al. [3] in order to determine the optimal formulation that balanced strength, insulation, and low weight. Jute fabric is used as reinforcement, greatly enhanced mechanical performance, according to Alcaraz et al. [4]. A gypsum–date palm fiber composite with enhanced mechanical, thermal, and water properties was made by Gallala et al. [5] using agricultural waste. Naiiri et al. [6] investigated the use of chemically treated doum fibers to reinforce gypsum mortars. They demonstrated how effectively the soda treatment enhanced both the fiber-matrix bonding and the thermo-mechanical behavior of these materials. Since their careful integration reduces bulk density by 17.16% and thermal conductivity by 26.24%, Khrissi et al. [7] confirmed the potential of date palm fibers as an insulating material for external cladding. Other studies such as those by Gencel et al. [8], Khalil et al. [9], and Benazzouk et al. [10] have shown that the gypsum's physical and mechanical characteristics matrix are significantly improved by the addition of particular industrial or agricultural waste. This creates new opportunities for residue recovery in sustainable building. Despite being widely available in desert and semi-desert areas, date palm fibers are still not widely used in gypsum composites. Up until now, the majority of research has concentrated on the mechanical characteristics or other kinds of plant reinforcements, frequently ignoring in-depth analyses of thermophysical characteristics and their modeling in practical application scenarios. We evaluated the viability of reinforcing with untreated Date-palm fibers in a matrix made of gypsum using numerical simulations and experiments as part of this approach. Our primary objective was to characterize these composite samples' thermophysical behavior according to their fiber percentage.This will be followed by simulating the thermal behavior of a wall incorporating these materials using COMSOL Multiphysics software. This work aims to demonstrate the viability of these materials as a sustainable and competitive thermal insulation solution, adapted to the needs of oasis areas.

Raw materials used

Plaster
Plaster is used extensively in the construction of buildings, not only for cladding walls and ceilings but also as a part of the fire protection system. It consists of dihydrated calcium sulfate ($CaSO_4·2H_2O$). Upon contact with water, it undergoes a setting process which generates heat and hardens the material, setting it apart from cement-based mortars. Natural gypsum is calcined at temperatures between 120 and 150°C to produce a partially dehydrated product (hemihydrate gypsum). The latter on mixing with water undergoes a well-defined chemical reaction Eq. 1 so that it can rehydrate and set.

$$2CaSO_4 . {}^1/_2 \; H_2O + 3H_2O \rightarrow 2CaSO_4 . 2H_2O \tag{1}$$

Emerging Research in Materials for Environment, and Civil Infrastructure - GeoME 5.5   Materials Research Forum LLC
Materials Research Proceedings 58 (2026) 54-61   https://doi.org/10.21741/9781644903933-8

Fibers from date palm trees

The saharat (the Saharan side) of Morocco, especially the Ziz and Drâa valleys are noted for their date palms. The waste from these types of trees, for example male palm fibers, is typically disregarded. In the present study, such unprocessed fibers were introduced in a gypsum binder as one of the reinforcement and formulating an indigenous eco-friendly material. Extracted from Drâa-Tafilalet region in Morocco, the fibres were cleaned and dried then cut to around 5 mm length segments, sieved, before being used as reinforcement into composite.

Sample preparation

The composite gypsum-ts fiber composites were fabricated using a approach based on the procedure reported by Kriker et al. [11] to acquire a homogenous distribution of the constituents. The moistened fibers were added within 5 min to the dry gypsum. Water was subsequently mixed at a liquid-to-gypsum ratio, 0.60, followed by rapid hand stirring to avoid early setting. Testing was carried out under fiber content, which ranged from 1% to 5%. The cast in a cylindrical steel molds (2 cm in height and 6 cm in diameter) made specifically for thermal characterization tests. A reference specimen without fibers was likewise prepared for comparison. The produced samples are illustrated in Fig. 1.

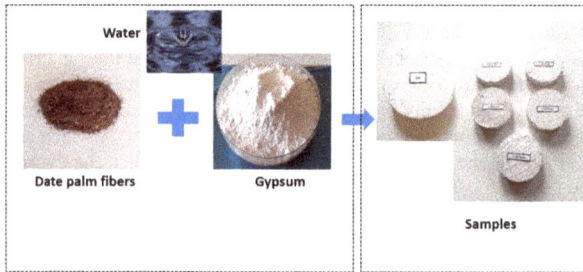

Fig. 1. Specimen preparation protocol

Experimental description of thermal properties measurement

The composite thermal characteristics, were simultaneously determined using the Hot Disk TPS 1500 system. This technique, compliant with ISO 22007-2 [12], The transient plane source approach serves as its foundation. It involves monitoring the heat response of a material exposed to a short-duration heat pulse. A double spiral-shaped sensor, that functions as a temperature sensor in addition to a heating element, is positioned between two symmetrical sections of the sample (see Fig. 2). Following a 15-minute stabilization period, an electric current is passed through the electrode chip to generate controlled heating. The temperature difference thus obtained is further analyzed to determine the material's thermal transport. The theoretical background of this technique was originally described by Gustafsson [13] and confirmed in subsequent studies [14 - 16].

Emerging Research in Materials for Environment, and Civil Infrastructure - GeoME 5.5   Materials Research Forum LLC
Materials Research Proceedings 58 (2026) 54-61   https://doi.org/10.21741/9781644903933-8

Fig. 2. Thermal properties testing device

Results and discussion

The thermal characteristics of the composites are improved as fibres derived from date palm trees are increasingly mixed with gypsum matrix. Thermal conductive decreases gradually from 0.522 W/m·K for the unfilled matrix to 0.394 W/m·K for the composite with 5% fiber, as shown in Table 1. This decrease demonstrates the porous character of the fibers, the heat cannot conduct within material. A decrease is also observed in thermal diffusivity, whic shows an increase in the thermal inertia of the matrix as well as a greater specific heat capacity considering it to have better capacity for energy storage ability. Effusivity also decreases, which is a further proof to the insulation capability of the material. The results show that our obtained values are lower than some which have been reported in previous studies [17-19] but higher than others [20-22] due to differences in formulation, fiber type, measurement methods, and experimental conditions. These biocomposites thus represent a high-performance, economical, and sustainable thermal solution suitable for energy-efficient buildings, particularly in warm climates.

Table 1. Thermal characteristics of the composites produced in this work

Specimen	Thermal conductivity [W/m. K]	Thermal diffusivity [m²/s]	Specific heat capacity Cp [J/kg. K]	Thermal effusivity [W.s$^{1/2}$/m². K]
Pure gypsum	0.522	5.60×10^{-7}	1090	695
Gypsum + 1 % fibers	0.508	5.35×10^{-7}	1110	691
Gypsum + 2 % fibers	0.479	5.10×10^{-7}	1130	680
Gypsum + 3% fibers	0.454	4.80×10^{-7}	1150	670
Gypsum + 4 % fibers	0.424	4.55×10^{-7}	1175	658
Gypsum + 5 % fibers	0.394	4.25×10^{-7}	1200	644

Application-Thermal Bridges in Building Construction

Energy conservation and sustainable practices are becoming increasingly important, necessitating solutions that protect the environment while reducing the energy demands of human activity. The construction industry is notorious for its high energy consumption and environmental impact, and the drive for sustainable growth promotes the introduction of creative alternatives while reducing the use of natural resources for construction. One innovation that supports the economy, energy conservation, and environmental preservation is the application of construction materials like plaster reinforced with date palm waste. Reducing heat leakage through structural joints becomes increasingly important as residential buildings become more energy-efficient, lowering thermal bridges and raising the thermal efficiency of the house. Custom thermal bridge designs reduce the likelihood of condensation and cold spots, allowing for energy efficiency requirements to be met

Emerging Research in Materials for Environment, and Civil Infrastructure - GeoME 5.5   Materials Research Forum LLC
Materials Research Proceedings 58 (2026) 54-61   https://doi.org/10.21741/9781644903933-8

without the need for additional renewable energy sources or other interventions. This improves the structural quality and reduces costs. A thermal bridge model that combines two- and three-dimensional designs is necessary for accurate heat transfer calculations, figuring out the overall thermal losses in a building or a particular area of it, and evaluating minimum surface temperatures to reduce the chance of condensation on surfaces. This entails defining the required thermal coefficients and interrelations, as well as the geometric structure of the model and thermal boundary conditions. Heat conduction through a building section more especially, between two levels is the subject of the analysis. The building is made up of four different materials, each with a different thermal conductivity. With the highest thermal conductivity (Fig. 3), the material used to divide the two levels plays a crucial role in thermal management at both internal (20°C) and external (0°C) temperature variations.

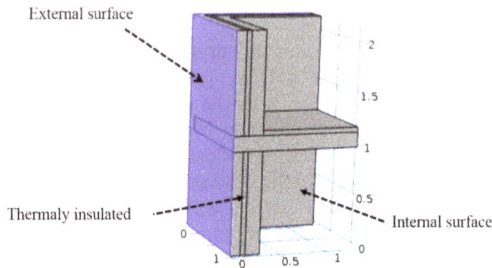

Fig. 3. Computational domain considered: thermally insulated, interior and exterior view

The thermal scenario is depicted in Fig. 4, where it is clear that the heat dissipation is greatest near the thermal bridge made by a horizontal piece at the wall. The thermal image shows heat migrating from the interior, represented by a colour transition of yellow (warm) to red (cool) tones across the slab of the balcony. This figure emphasizes the advantage of a load-bearing insulation member in reducing thermal losses and maintaining the integrity of the insulating material. An in-depth analysis of this figure shows that the selection of gypsum and date palm tree fibers greatly affect thermal responses of the composite in a building. The use of these materials ensures that, in various scenarios, the slab retains a higher temperature. Although this thermal impact is also observed at the juncture of two vertical walls, it is mitigated by an internal insulation layer.

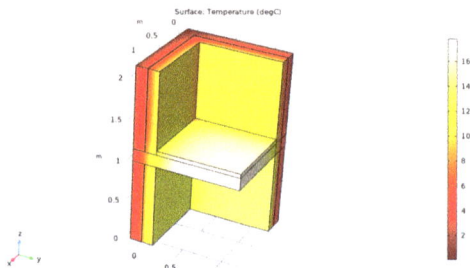

Fig. 4. Temperature distribution

Emerging Research in Materials for Environment, and Civil Infrastructure - GeoME 5.5   Materials Research Forum LLC
Materials Research Proceedings 58 (2026) 54-61   https://doi.org/10.21741/9781644903933-8

Fig. 5 we show the lowest and highest surface temperature at the bounds. Similarly, one must also use a finite element analysis to obtain the minimum temperature at the thermal bridge given node design and building materials used.

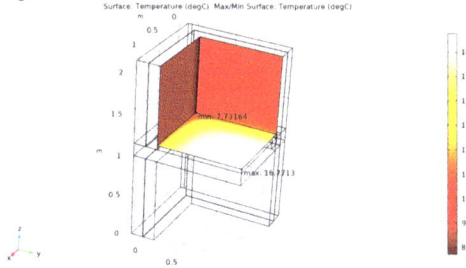

Fig. 5. Minimum and maximum temperatures

Table 2 enumerates the results of research that report on the heat transfer across surfaces of a heated space during its coldest week of the year, along with corresponding extreme high and low surface temperatures. It is emphasized that even by using less thermally conductive materials, the heat dissipation from the actuated zone can be greatly reduced.

Table 2. Maximum temperature, Maximum temperature, and Heat flux in gypsum reinforced with date palm fibers at the various percentage

Sample	Minimum temperature	Maximum temperature	Heat flux
Pure gypsum	7.7316 °C	16.7713 °C	46.254 W
Gypsum + 1% fibers	7.7334 °C	16.7714 °C	46.194 W
Gypsum+ 2% fibers	7.7337 °C	16.7714 °C	46.124 W
Gypsum + 3% fibers	7.7375 °C	16.7714 °C	46.074 W
Gypsum + 4% fibers	7.7400°C	16.7715 °C	46.014 W
Gypsum + 5% fibers	7.7413°C	16.7715 °C	45.954 W

In this study, we conducted experimental analyses of the plaster and date palm fiber mixture. The aim was to integrate these results into a numerical model that simulates the dynamic transfer of heat conduction in a conventional construction framework, focusing on a three-dimensional (3D) representation of the relevant variables. We looked into the consequences of changing ratios of date palm fiber to plaster on the thermal properties of the 3D structure. Heat loss in heated spaces can be considerably decreased by about 75% when advanced composites made of date palm fibers and plaster are used in construction. In addition to lowering greenhouse gas emissions, this reduction results in considerable heating cost savings. Because these materials have higher mean radiant and operating temperatures, they improve thermal and moisture comfort in buildings. These cutting-edge materials also stop water from condensing around structural components. In

Emerging Research in Materials for Environment, and Civil Infrastructure - GeoME 5.5   Materials Research Forum LLC
Materials Research Proceedings 58 (2026) 54-61   https://doi.org/10.21741/9781644903933-8

general, the construction sector gains from the use of date palm fiber and gypsum composites since they lower greenhouse gas emissions and increase energy efficiency in structural applications.

Conclusion

The current study showed how well palm tree fibers work as a thermal enhancer in composites made of gypsum. Fibers from date palms are a plentiful natural resource that can be found locally. The experimental results showed that gradually increasing the fiber content increased the heat capacity while also causing a discernible reduction in diffusivity and thermal conductivity. These alterations demonstrate that the material's overall insulating effectiveness has increased. Numerical modelling based on the experimental data confirmed these findings, showing that heat transfer through thermal bridges was reduced, leading to greater thermal comfort indoors. In comparison to conventional insulation products, the resulting biocomposites provide a good balance between thermal performance, cost, longevity, and environmental sustainability. Especially in hotter climates, their application in external wall systems can reduce greenhouse gas emissions and building energy consumption. These materials present a practical and eco-friendly alternative to sustainable construction, and they hold great promise for application in green housing initiatives, especially in Morocco's oasis regions.

References

[1] T. Alioua, et al., Sensitivity analysis of transient heat and moisture transfer in a bio-based date palm concrete wall, Building and Environment 202 (2021) 108019.

[2] M.M. Khenfer, P.P. Morlier, Plâtres renforcés de fibres cellulosiques, Materials and Structures 32 (1999) 52–58. https://doi.org/10.1007/BF02480412

[3] Y. Maaloufa, et al., Thermal and mechanical behavior of the plaster reinforced by fiber alpha or granular cork, Civil Engineering and Technology 8 (2017) 1026–1040.

[4] J.S. Alcaraz, et al., Mechanical properties of plaster reinforced with jute fabrics, Composites Part B: Engineering 178 (2019) 107390. https://doi.org/10.1016/j.compositesb.2019.107390

[5] W. Gallala, et al., Production of low-cost biocomposite made of palm fibers waste and gypsum plaster, Proceedings of the International Conference on Contaminated Environment 36 (2020) 475–483.

[6] F. Naiiri, et al., The effect of doum palm fibers on the mechanical and thermal properties of gypsum mortar, Journal of Composite Materials 53 (2019) 2641–2659.

[7] Y. Khrissi, A. Tilioua, N. Laaroussi, Thermal characterization of a new bio-composite building material based on gypsum and date palm fiber, Materials Research Proceedings 40 (2024) 55–63. https://doi.org/10.21741/9781644903117-6

[8] O. Gencel, et al., Properties of gypsum composites containing vermiculite and polypropylene fibers: Numerical and experimental results, Energy and Buildings 70 (2014) 135–144. https://doi.org/10.1016/j.enbuild.2013.11.047

[9] A.A. Khalil, et al., Effect of some waste additives on the physical and mechanical properties of gypsum plaster composites, Construction and Building Materials 68 (2014) 580–586. https://doi.org/10.1016/j.conbuildmat.2014.06.081

[10] A. Benazzouk, et al., Thermal conductivity of cement composites containing rubber waste particles: Experimental study and modelling, Construction and Building Materials 22 (2008) 573–579. https://doi.org/10.1016/j.conbuildmat.2006.11.011

[11] A. Kriker, et al., Mechanical properties of date palm fibres and concrete reinforced with date palm fibres in hot-dry climate, Cement and Concrete Composites 27 (2005) 554–564.

[12] ISO 22007-2, Plastics — Determination of thermal conductivity and thermal diffusivity — Part 2: Transient plane heat source (hot disc) method, (2015).

[13] S.E. Gustafsson, Transient plane source technique for thermal conductivity and thermal diffusivity measurements of solid materials, Review of Scientific Instruments 62 (1991) 797–804.

[14] T. Log, S.E. Gustafsson, Transient Plane Source (TPS) Technique for Measuring Thermal Transport Properties of Building Materials, Fire and Materials 19 (1995) 43–49. https://doi.org/10.1002/fam.810190107

[15] D. Salmon, Thermal conductivity of insulations using guarded hot plates, including recent developments and sources of reference materials, Measurement Science and Technology 12 (2001) R107–R116. https://doi.org/10.1088/0957-0233/12/12/201

[16] S. Malinarič, Parameter estimation in dynamic plane source method, Measurement Science and Technology 15 (2004) 807–813. https://doi.org/ 10.1088/0957-0233/15/5/005

[17] M. Rachedi, A. Kriker, Investigation of the mechanical and thermal characteristics of an eco-insulating material made of plaster and date palm fibers, Journal of Civil Engineering 16 (2021) 55–66.

[18] A. Djoudi, et al., Effect of the addition of date palm fibers on thermal properties of plaster concrete: Experimental study and modeling, Journal of Adhesion Science and Technology 28 (2014) 2100–2111. https://doi.org/10.1080/01694243.2014.948363

[19] C.C. Pinto, R.F. Carvalho, Thermal performance evaluation of a low-cost housing ceiling prototype made with gypsum and sisal fibre panels, IOP Conference Series: Earth and Environmental Science 296 (2019) 012015. 10.1088/1755-1315/296/1/012015

[20] N. Vavřínová, et al., Research of mechanical and thermal properties of composite material based on gypsum and straw, Journal of Renewable Materials 10 (2022) 1859–1873. https://doi.org/10.32604/jrm.2022.018908

[21] A. Braieka, et al., Estimation of the thermophysical properties of date palm fibers/gypsum composite for use as insulating materials in building, Energy and Buildings 140 (2017) 268–279. https://doi.org/10.1016/j.enbuild.2017.02.001

[22] V. Guna, et al., Wool and coir fiber reinforced gypsum ceiling tiles with enhanced stability and acoustic and thermal resistance, Journal of Building Engineering 41 (2021) 102433. https://doi.org/10.1016/j.jobe.2021.102433

Emerging Research in Materials for Environment, and Civil Infrastructure - GeoME 5.5   Materials Research Forum LLC
Materials Research Proceedings 58 (2026) 62-69   https://doi.org/10.21741/9781644903933-9

3D concrete printing technology:
Progress and prospects for sustainable construction

Nada Oulad Moussa[1,a] *, Mohamed El Haim[1,b] and Loubaba Rida[2,3,c]

[1]Civil Engineering, Mechanics and Energy Research Team: Modeling and Experimentation (GC2-ME), National School of Applied Sciences of Al Hoceima, Abdelmalek Essaadi University, Al Hoceima, Morocco

[2]Laboratory of Systems, Control, and Decision (LSCD), New Science School of Engineering (ENSI), Tangier, Morocco

[3]Mechanical and Civil Engineering Laboratory, FST of Tangier, Abdelmalek Essaadi University, Tangier, Morocco

[a]nadaom.professionnel@gmail.com, [b]melhaim@uae.ac.ma, [c]Loubaba.rida@gmail.com

Keywords: 3D Printed Concrete, Sustainable Construction, Low-Carbon Concrete, Eco-Friendly Materials, Digital Fabrication

Abstract. The evolution of the 3D concrete printing technology is a breakthrough in the history of construction technologies. Three-dimensional concrete printing can streamline building processes, reducing material waste and labor time while boosting overall productivity. The technology has many environmental sustainability potentials by saving material waste, labor dependence, formwork disassembly, and energy-efficient building materials. This study highlights recent progress in 3DCP materials and applications, along with the associated challenges and prospects. The paper emphasizes the mechanical behavior, durability, and long service life of low-carbon 3D-printed concrete and significant problems such as printability, long durability, and interlayer adhesion. To address sustainability challenges, we talk about eco-friendly alternatives such as alkali-activated geopolymers, calcium-sulfo-aluminate, and limestone-calcined-clay cements. Such materials lower the carbon footprint of cementitious composites and improve the long-term strength performance of building elements. 3DPC represents a significant shift towards more sustainable production and anticipates further advancements in this revolutionized construction.

Introduction

The construction industry is crucial for modern society, as it creates the necessary infrastructure and drives economic growth. Nevertheless, the industry is responsible for most of the environmental issues, generating around 40% of the world's energy consumption, 28% of the production of greenhouse gas, and also around 40% of all solid waste produced [1]. The most common building material, concrete, is among the largest environmentally unfriendly materials because it is responsible for the largest carbon footprints due to the production of cement, amounting to about 5–8% of global CO_2 emissions [2]. Traditional construction methods are labor-intensive, time-consuming, and inefficient, often resulting in material waste, project delays, and overbudgeting [3].

The three-dimensional concrete printing refers to a new method that uses digital fabrication technologies to build structures in incremental layers with great accuracy [4],[5]. It minimizes waste, reduces human error, and enhances the efficiency of time and cost [6], [7]. Besides this, 3DCP contributes to sustainability by incorporating environmentally friendly materials, like alkali-activated geopolymers, calcium-sulfo-aluminate, and limestone-calcined-clay cements [8] and assures the consumption of resources only in places where they are required. These binders also permit large clinker substitution and usually surpass OPC in several ways. For instance, geopolymers and LC3 cements are outlined as competing, and even surpassing, OPC mixes in hardened strength, at the same time, cut embodied CO_2 emissions by roughly one-third to one-half 30–50% [9].

Addressing environmental concerns and improving efficiency in construction, 3DCP offers a revolutionary concept of building that will meet global sustainable development goals and lead to cleaner, faster, and adaptive construction [3],[10].

Binder materials

3DCP technology requires concrete that must possess some extrudability and buildability properties, for which the material selection forming the concrete must possess some specific requirements [11]. The choice of material must also ensure correct flow in the pump system, smooth extrusion from the nozzle, and appropriate structural integrity after deposition [11],[12]. Other than these, rheological and mechanical performances of the 3D-printed concrete in fresh or in the hardened state, in turn, include the main criteria in selecting appropriate materials [12].

The next section reveals the most prevalent low-carbon raw materials for cement-based building material manufacturing.

Geopolymer Binders

Geopolymers are a type of alkali-activated material, which are extensively researched in 3DPC due to their capability of replacing ordinary Portland cement and reducing of concrete's embodied CO_2 by 30–80 % [13],[14]. They also possess potential for CO2 footprint reduction related to the production of traditional concrete [13],[15].

Geopolymers are produced from aluminosilicate ashes, slags, or metakaolins, which undergo chemical activation through the addition of alkaline chemicals, such as sodium metasilicate ($NaNa_2SiO_3$) [13], [14],[16]. It eliminates dangerous liquid activators used in two-part geopolymers (liquid activator + precursor) to deal with [13]. The utilization of solid activators has the potential to improve the fresh characteristics of one-part geopolymers, including their extrusion ease, structural integrity, and printing quality, when juxtaposed with two-part systems. Nevertheless, this enhancement frequently incurs a trade-off, resulting in diminished open time and decreased flowability [17],[18].

The rheology can be adjusted by mix design, incorporating blast furnace slag (GGBS) and fine sand enhances static yield stress (upgrading build-up capacity) but decreases consistency coefficient [19]. Through optimization of the geopolymer mix, we can obtain elevated yield stress for stackability, yet not too high viscosity, exhibiting good printability [13]. Also, the incorporation of fibers (e.g., 0.6% PE fiber) improved flexural toughness by about 28% [20].

Moreover, geopolymers also exhibit good durability (limited shrinkage and strong chemical resistance) [13].

Multiple studies [21], [22], [23] have examined how variations in binder composition affect the behavior and performance of geopolymer-based materials, with a summary of their findings provided in Table 1.

Table 1: various 3D-printed geopolymer formulations

Binder composition	Key mechanical & rheological findings	references
Metakaolin-based geopolymer: Mixe1 = 100% metakaolin Mixe2 = 95% metakaolin + 5% GGBFS (with NaOH + Na₂SiO₃ activator).	5% slag lowered the setting time significantly (from 17h to 4h) and marginally raised the static/dynamic yield stress, increasing the buildable lifts from 27 to 42. At 28 d, the M2 was more robust: the compressive strength was approximately 11–21% greater, and flexural strength was approximately 28–33% greater compared to the M1. Slag	[23]

	lowered the drying shrinkage and the water absorption (better durability index).	
Fly ash + quartz sand (with NaOH + Na₂SiO₃ activator).	Incorporating ~10% GGBFS significantly increased paste viscosity and yield stress (better thixotropy) in FA-sand geopolymer.	[22]
50% high-Ca fly ash + 50% GGBFS (by binder mass); solid activators (dry NaOH/Na₂SiO₃ incorporated); water/binder 0.28–0.36; borax (Na₂B₄O₇·10H₂O) retarder 0–6% of binder.	A slump range of 15–30 mm, with a slump flow of 210–240 mm, typically gives mixes that not only are extrudable but also retain their shape after placement. For open time >20 min, required ≥2–4% borax with W/B >0.34. 3-day compressive strength was approximately 16.7 MPa for W/B=0.32 with 2% borax (compliant with 3D-print requirements). Strength dropped sharply when borax >4%	[21]

Calcium Sulfoaluminate (CSA) cement
Relatively to ordinary Portland cement (OPC), calcium sulfoaluminate (CSA) presents a more environmentally friendly solution, with far fewer CO_2 emissions in the production of clinker. [24] The CSA clinkers may also be generated through sintering of aluminum-containing minerals, calcium sulfate, as well as calcium carbonate in the range of about 1250 °C, which is about 200 °C lower than the typical temperature utilized for OPC clinker manufacture [25]. CSA clinker cement can also be produced from industrial residuals, such as iron-making sludge, residual red oxide, marble processing residues, and aluminum-containing ore[25],[26].

CSA also accelerates setting and early strength, desirable for achieving high printing speed, but which must be restrained carefully (typically with further plaster or retarders) lest the plaster unexpectedly flash sets [27]. In one sample, a blend with 7% CSA and 93% OPC achieved approximately 88 MPa 28-day compressive strength when poured, and ~79 MPa when 3D-printed [28].

Mechanically, the CSA-based 3D-printed specimens show greater flexural strength compared with the normal printed OPC mixes [29]. For example, the flexural strength was 0.88 MPa for the OPC, but increased swiftly with respect to the addition of CSA, and the optimum strength was found as 2.88 MPa at 10 wt% CSA, with the highest volume formation of ettringite [29],[30].

Limestone Calcined Clay Cement (LC3)
Limestone-calcined clay cements (LC3) are regarded as viable, resilient, and environmentally sustainable substitutes for conventional cement [31]. This mixture lowers the clinker content by about 50% without sacrificing performance, equal to that of OPC [32], [33]. The main reactive components in calcined clay are metakaolin (MK), which reacts with the calcium hydroxide formed during the cement hydration process to form calcium–aluminosilicate–hydrate molecules [31],[34],[35]. Simultaneously, hemi- and mono-carboaluminate phases form because of reactions between alumina species present in the pore solution and calcite (supplied by the limestone) [31], [34].

LC3 continues to offer a viable solution for carbon emission minimization in the process of 3D concrete printing, with adequate mechanical strength in association with a higher setting rate in the mixes with a lower content of clinker [36].

A summary is presented in Table 2 showing that several studies have investigated the influence of binder composition on the performance of Limestone calcined clay cements.

Emerging Research in Materials for Environment, and Civil Infrastructure - GeoME 5.5   Materials Research Forum LLC
Materials Research Proceedings 58 (2026) 62-69   https://doi.org/10.21741/9781644903933-9

Table 2: variation in the proportions of cement, calcination-clay, and limestone (and occasional fibres or admixtures) affects fresh-state rheology, buildability, and hardened-state strength for 3-D-printed concrete.

Binder composition	Main performance outcomes for 3D-CP	references
Binary mix: 20 % calcined clay (CC) + 80 % cement + lime filler (no limestone) • **Ternary mix: 30 % CC + 15 % limestone filler (LF) + 55 % cement** • **2 % 6-mm steel fibers added for reinforcement**	• Both mixes showed high packing density and water-retention, giving good extrudability. • 2 % steel fibre raised 28-day compressive and 28-day flexural strength to the highest values reported for the study.	[33]
Typical formulation: 50 % OPC replaced by a blend of 30 % calcined clay + 15 % limestone (by binder mass) + 5 % gypsum (the "LC3" blend).	Provides the same 30 MPa 28-day compressive strength as a reference OPC mix while cutting the climate-change (global-warming-potential) score by 36 % (QC) to 46 % (FR). • Higher static yield stress and early-age stiffness give good shape-retention for 3D printing.	[36]
Portland cement substituted by >60 % total of calcined clay (CC) + limestone filler (LF) (e.g., 70 % CC + 15 % LF).	Slump and flowability drop markedly, but buildability improves strongly because water-film thickness (WFT) is reduced. • A higher CC content promotes faster development of early stiffness and an increase in total specific surface area during the first 3 hours. • Dilution effect lowers compressive strength proportionally to the replacement level.	[34]

Challenges and Opportunities

Rheology and Printability

Identification of the optimal trade-off between fluidity (for pumping/extruding) and static yield stress (for buildability) is paramount [37]. Low-carbon binders exhibit continuous variations in fresh rheology [37]. When nano- or micro-fibrillated cellulose is incorporated into geopolymer formulations, yield stress and plastic viscosity increase, with a corresponding large increase in thixotropy. This increases the build strength of the material, enabling it to withstand as many as 32 deposited layers before structural failure [38]. For the LC3 system, reducing the superplasticizer-to-viscosity-modifying-agent (SP/VMA) ratio from 2.6 to 1.3 gives higher static yield stress and increased thixotropic behavior, with stiffer filaments that can sustain an optimal build height of approximately 40 layers [33],[39]. Printability forecasting relationships and inline rheometers are becoming universal methods to express extrudability, buildability, and shape retention numerically [32].

Interlayer Bonding and Anisotropy

The weak interlayer interface is the main disadvantage in 3DPC. Anisotropy implies that the tensile/flexural strength normal to the layers is only about 50–70% that in the in-lay direction. Addition of short, hard fibers (e.g., steel) has much reduced the effects of anisotropy: fiber bridges between the layers suppress interface cracks; too much fiber, though, can cause nozzle clogging and non-uniform bonding [32], [40]. The process and printing parameters of the material (lower layer height, vibration, or surface activation) also have effects on the bond. Prior continuous reinforcement (e.g., rebar or filament winding) is still an open avenue for future work [32], [40].

Setting and Strength Development

CSA concretes have quick sets (minutes to hours) that accelerate initial strength development but can induce nozzle clogging and cracking [37]. Fine control over gypsum/extenders can be used to control CSA sets. Geopolymers tend towards slow initial setting; addition of small amounts of slag or use of a one-part mix can increase setting significantly [37]. Illustration of the MK-geopolymer work showed that the addition of slag lowered the setting time from 17 to 4 h [37].

Conclusion

Overall, low-carbon binders have high potential for 3D-printed concrete. Geopolymers and LC3 cements can offer desired strength and reduce CO_2 emissions considerably. CSA cements have a very rapid set and appropriate initial strength at low-carbon cost. However, all low-carbon systems must cope with the inherent anisotropy and porosity of layer printing. Fiber reinforcement and optimal mix design are efficient ways to enhance the strength and durability between layers. Life-cycle analyses invariably favor high-SCM or geopolymer mixes on sustainability, although the full benefit is source material and energy dependent.

Key areas that need further exploration include standardization of test methods on 3DPC, scale-up production on sustainable binder, and measurement of long-term field performance.

In general, the use of low-carbon binders in 3D printing is compatible with global decarbonization, though it is subject to multi-faceted advancements in material engineering, printing technology, and environmental measurement.

Reference

[1] I. Agustí-Juan and G. Habert, "Environmental design guidelines for digital fabrication," *J Clean Prod*, vol. 142, pp. 2780–2791, Jan. 2017. https://doi.org/10.1016/J.JCLEPRO.2016.10.190

[2] S. Sbahieh, M. Zaher Serdar, and S. G. Al-Ghamdi, "Decarbonization strategies of building materials used in the construction industry," *Mater Today Proc*, Sep. 2023. https://doi.org/10.1016/J.MATPR.2023.08.346

[3] T. Tabassum and A. Ahmad Mir, "A review of 3d printing technology-the future of sustainable construction," *Mater Today Proc*, vol. 93, pp. 408–414, Jan. 2023. https://doi.org/10.1016/J.MATPR.2023.08.013

[4] S. A. M. Tofail, E. P. Koumoulos, A. Bandyopadhyay, S. Bose, L. O'Donoghue, and C. Charitidis, "Additive manufacturing: scientific and technological challenges, market uptake and opportunities," *Materials Today*, vol. 21, no. 1, pp. 22–37, Jan. 2018. https://doi.org/10.1016/J.MATTOD.2017.07.001

[5] F. Craveiro, S. Nazarian, H. Bartolo, P. J. Bartolo, and J. Pinto Duarte, "An automated system for 3D printing functionally graded concrete-based materials," *Addit Manuf*, vol. 33, p. 101146, May 2020. https://doi.org/10.1016/J.ADDMA.2020.101146

[6] M. Adaloudis and J. Bonnin Roca, "Sustainability tradeoffs in the adoption of 3D Concrete Printing in the construction industry," *J Clean Prod*, vol. 307, p. 127201, Jul. 2021. https://doi.org/10.1016/J.JCLEPRO.2021.127201

[7] V. Mechtcherine, V. N. Nerella, F. Will, M. Näther, J. Otto, and M. Krause, "Large-scale digital concrete construction – CONPrint3D concept for on-site, monolithic 3D-printing," *Autom Constr*, vol. 107, p. 102933, Nov. 2019. https://doi.org/10.1016/J.AUTCON.2019.102933

[8] S. Bhattacherjee *et al.*, "Sustainable materials for 3D concrete printing," *Cem Concr Compos*, vol. 122, p. 104156, Sep. 2021. https://doi.org/10.1016/J.CEMCONCOMP.2021.104156

[9] Y. A. Al-Noaimat *et al.*, "3D printing of limestone-calcined clay cement: A review of its potential implementation in the construction industry," *Results in Engineering*, vol. 18, p. 101115, Jun. 2023. https://doi.org/10.1016/J.RINENG.2023.101115

[10] A. H. Alami, A. G. Olabi, M. Ayoub, H. Aljaghoub, S. Alasad, and M. A. Abdelkareem, "3D Concrete Printing: Recent Progress, Applications, Challenges, and Role in Achieving Sustainable Development Goals," *Buildings 2023, Vol. 13, Page 924*, vol. 13, no. 4, p. 924, Mar. 2023. https://doi.org/10.3390/BUILDINGS13040924

[11] P. Panchal and M. S. Choi, "A review on effect of natural fibers to mitigate CO2 footprint and enhance engineering properties of 3D printing concrete," *Journal of Building Engineering*, vol. 111, p. 113562, Oct. 2025. https://doi.org/10.1016/J.JOBE.2025.113562

[12] Z. Malaeb, F. AlSakka, and F. Hamzeh, "3D Concrete Printing: Machine Design, Mix Proportioning, and Mix Comparison Between Different Machine Setups," *3D Concrete Printing Technology: Construction and Building Applications*, pp. 115–136, Jan. 2019. https://doi.org/10.1016/B978-0-12-815481-6.00006-3

[13] Y. A. Al-Noaimat, S. H. Ghaffar, M. Chougan, and M. J. Al-Kheetan, "A review of 3D printing low-carbon concrete with one-part geopolymer: Engineering, environmental and economic feasibility," *Case Studies in Construction Materials*, vol. 18, p. e01818, Jul. 2023. https://doi.org/10.1016/J.CSCM.2022.E01818

[14] S. H. Bong, M. Xia, B. Nematollahi, and C. Shi, "Ambient temperature cured 'just-add-water' geopolymer for 3D concrete printing applications," *Cem Concr Compos*, vol. 121, p. 104060, Aug. 2021. https://doi.org/10.1016/J.CEMCONCOMP.2021.104060

[15] L. Rida, K. Bazzar, and A. H. Alaoui, "High-Volume Fly Ash Mortar Solution for Sustainable Development," *Advances in Intelligent Systems and Computing*, vol. 1104 AISC, pp. 386–395, 2020. https://doi.org/10.1007/978-3-030-36671-1_33

[16] N. B. Singh and B. Middendorf, "Geopolymers as an alternative to Portland cement: An overview," *Constr Build Mater*, vol. 237, p. 117455, Mar. 2020. https://doi.org/10.1016/J.CONBUILDMAT.2019.117455

[17] S. Muthukrishnan, S. Ramakrishnan, and J. Sanjayan, "Effect of alkali reactions on the rheology of one-part 3D printable geopolymer concrete," *Cem Concr Compos*, vol. 116, p. 103899, Feb. 2021. https://doi.org/10.1016/J.CEMCONCOMP.2020.103899

[18] M. H. Raza, R. Y. Zhong, and M. Khan, "Recent advances and productivity analysis of 3D printed geopolymers," *Addit Manuf*, vol. 52, p. 102685, Apr. 2022. https://doi.org/10.1016/J.ADDMA.2022.102685

[19] B. Panda, G. B. Singh, C. Unluer, and M. J. Tan, "Synthesis and characterization of one-part geopolymers for extrusion based 3D concrete printing," *J Clean Prod*, vol. 220, pp. 610–619, May 2019. https://doi.org/10.1016/J.JCLEPRO.2019.02.185

[20] X. Zhu, J. Wang, M. Yang, J. Xiao, Y. Zhang, and F. A. Gilabert, "Performance modulation and optimization of PE fiber reinforced 3D-printed geopolymer," *Constr Build Mater*, vol. 429, p. 136449, May 2024. https://doi.org/10.1016/J.CONBUILDMAT.2024.136449

[21] M. V. Tran, T. H. Vu, and T. H. Y. Nguyen, "Simplified assessment for one-part 3D-printable geopolymer concrete based on slump and slump flow measurements," *Case Studies in Construction Materials*, vol. 18, p. e01889, Jul. 2023. https://doi.org/10.1016/J.CSCM.2023.E01889

[22] X. Guo, J. Yang, and G. Xiong, "Influence of supplementary cementitious materials on rheological properties of 3D printed fly ash based geopolymer," *Cem Concr Compos*, vol. 114, p. 103820, Nov. 2020. https://doi.org/10.1016/J.CEMCONCOMP.2020.103820

[23] M. B. Jaji, G. P. A. G. van Zijl, and A. J. Babafemi, "Durability and pore structure of metakaolin-based 3D printed geopolymer concrete," *Constr Build Mater*, vol. 422, p. 135847, Apr. 2024. https://doi.org/10.1016/J.CONBUILDMAT.2024.135847

[24] M. A. G. Aranda and A. G. De la Torre, "Sulfoaluminate cement," *Eco-Efficient Concrete*, pp. 488–522, Jan. 2013. https://doi.org/10.1533/9780857098993.4.488

[25] G. A. N. Yanze *et al.*, "Development of calcium sulfoaluminate cements from rich-alumina bauxite and marble wastes: Physicochemical and microstructural characterization," *International Journal of Ceramic Engineering & Science*, vol. 6, no. 3, p. e10216, May 2024. https://doi.org/10.1002/CES2.10216

[26] J. Park, J. Seo, S. Park, A. Cho, and H. K. Lee, "Phase profiling of carbonation-cured calcium sulfoaluminate cement," *Cem Concr Res*, vol. 189, p. 107776, Mar. 2025. https://doi.org/10.1016/J.CEMCONRES.2024.107776

[27] M. K. Mohan, A. V. Rahul, G. De Schutter, and K. Van Tittelboom, "Early age hydration, rheology and pumping characteristics of CSA cement-based 3D printable concrete," *Constr Build Mater*, vol. 275, p. 122136, Mar. 2021. https://doi.org/10.1016/J.CONBUILDMAT.2020.122136

[28] N. Khalil, G. Aouad, K. El Cheikh, and S. Rémond, "Use of calcium sulfoaluminate cements for setting control of 3D-printing mortars," *Constr Build Mater*, vol. 157, pp. 382–391, Dec. 2017. https://doi.org/10.1016/J.CONBUILDMAT.2017.09.109

[29] S. Kim, T. Kim, B. Kim, H. dae Kim, J. Kim, and H. Lee, "Early hydration and hardening of OPC-CSA blends for cementitious structure of 3D printing," *Advances in Applied Ceramics*, vol. 119, no. 7, pp. 393–397, Oct. 2020. https://doi.org/10.1080/17436753.2020.1777505

[30] S. Kim, T. Kim, B. Kim, H. dae Kim, J. Kim, and H. Lee, "Early hydration and hardening of OPC-CSA blends for cementitious structure of 3D printing," *Advances in Applied Ceramics*, vol. 119, no. 7, pp. 393–397, Oct. 2020. https://doi.org/10.1080/17436753.2020.1777505

[31] K. Scrivener, F. Martirena, S. Bishnoi, and S. Maity, "Calcined clay limestone cements (LC3)," *Cem Concr Res*, vol. 114, pp. 49–56, Dec. 2018. https://doi.org/10.1016/J.CEMCONRES.2017.08.017

[32] K. A. Ibrahim, G. P. A. G. van Zijl, and A. J. Babafemi, "Influence of limestone calcined clay cement on properties of 3D printed concrete for sustainable construction," *Journal of Building Engineering*, vol. 69, p. 106186, Jun. 2023. https://doi.org/10.1016/J.JOBE.2023.106186

[33] H. ; Li, J. ; Wei, K. H. Khayat, H. Li, J. Wei, and K. H. Khayat, "3D Printing of Fiber-Reinforced Calcined Clay-Limestone-Based Cementitious Materials: From Mixture Design to Printability Evaluation," *Buildings 2024, Vol. 14, Page 1666*, vol. 14, no. 6, p. 1666, Jun. 2024. https://doi.org/10.3390/BUILDINGS14061666

[34] Y. Chen, S. He, Y. Zhang, Z. Wan, O. Çopuroğlu, and E. Schlangen, "3D printing of calcined clay-limestone-based cementitious materials," *Cem Concr Res*, vol. 149, p. 106553, Nov. 2021. https://doi.org/10.1016/J.CEMCONRES.2021.106553

[35] Y. Tarhan, İ. H. Tarhan, and R. Şahin, "Comprehensive Review of Binder Matrices in 3D Printing Construction: Rheological Perspectives," *Buildings 2025, Vol. 15, Page 75*, vol. 15, no. 1, p. 75, Dec. 2024. https://doi.org/10.3390/BUILDINGS15010075

[36] W. Jin, C. Roux, C. Ouellet-Plamondon, and J. F. Caron, "Life cycle assessment of limestone calcined clay concrete: Potential for low-carbon 3D printing," *Sustainable Materials and Technologies*, vol. 41, p. e01119, Sep. 2024. https://doi.org/10.1016/J.SUSMAT.2024.E01119

[37] M. B. Jaji, G. P. A. G. van Zijl, and A. J. Babafemi, "Durability and pore structure of metakaolin-based 3D printed geopolymer concrete," *Constr Build Mater*, vol. 422, p. 135847, Apr. 2024. https://doi.org/10.1016/J.CONBUILDMAT.2024.135847

[38] Y. Chen *et al.*, "Rheology control and shrinkage mitigation of 3D printed geopolymer concrete using nanocellulose and magnesium oxide," *Constr Build Mater*, vol. 429, p. 136421, May 2024. https://doi.org/10.1016/J.CONBUILDMAT.2024.136421

[39] H. ; Li, J. ; Wei, K. H. Khayat, H. Li, J. Wei, and K. H. Khayat, "3D Printing of Fiber-Reinforced Calcined Clay-Limestone-Based Cementitious Materials: From Mixture Design to Printability Evaluation," *Buildings 2024, Vol. 14, Page 1666*, vol. 14, no. 6, p. 1666, Jun. 2024. https://doi.org/10.3390/BUILDINGS14061666

[40] M. Salmi, O. Sucharda, J. Akmal, S. Surehali, A. Tripathi, and N. Neithalath, "Anisotropy in Additively Manufactured Concrete Specimens under Compressive Loading—Quantification of the Effects of Layer Height and Fiber Reinforcement," *Materials 2023, Vol. 16, Page 5488*, vol. 16, no. 15, p. 5488, Aug. 2023. https://doi.org/10.3390/MA16155488

Emerging Research in Materials for Environment, and Civil Infrastructure - GeoME 5.5   Materials Research Forum LLC
Materials Research Proceedings 58 (2026) 70-77   https://doi.org/10.21741/9781644903933-10

Experimental and simulation study of the mechanical properties of eco composite materials reinforced with multiple fibers

Asmae OUROUI[1,a] *, Youssef BIBRIDNE[2,b], Najma LAAROUSSI[1,c],
Mohammed AIT EL FQIH[2,d] and Mohammed ELWAZNA[3,e]

[1]Mohammed V University in Rabat in Rabat, Material, Energy and Acoustics Team (MEAT), EST Salé, Morocco

[2]Laboratory of Artificial Intelligence & Complex Systems Engineering (AICSE), ENSAM, Hassan II University of Casablanca, Morocco

[3]Technical Manager of R&D and Innovation Programs, IRESEN

[a]asmae.ouroui@research.emi.ac.ma, [b]youssefbibrden@gmail.com,
[c]najma.laaroussi@est.um5.ac.ma, [d]m.aitelfqih@gmail.com, [e]elwazna@iresen.org

Keywords: Epoxy Matrix, Fibers, Hardness Test, Tensile Test, Mechanical Proprieties, Simulation

Abstract. The study focus in improving the performance of our composite made of an epoxy thermosetting matrix reinforced with jute, tire, and glass fibers. The composite is fabricated using manual contact molding at room temperature without applying pressure, followed by hardening and finishing using laser cutting technology. Specimens are prepared according to ASTM D638 standards. Abaqus is used to simulate the mechanical behavior using the Finite Element Method (FEM) using various loading conditions. This digital approach enables precise predictions of material performance, reducing the need for extensive experimental testing. Also, three dimensional modelling of the composite layers is completed in CATIA to ensure that the fibre thickness and orientation are accurately modelled. For six of the specimens investigated, the internal defects such as voids and microcracks and fibre orientation were examined using microscopy methods and the analysis of digital images. Tensile and hardness tests were conducted for measurement. The measured hardness values are 74.58 (Shore A) and 39.54 (Shore D). Microscopic analysis provides additional detailed information regarding internal structure and possible defects through visualising the bi-directional, planar perpendicular fibre arrangement.

Introduction

There are broadly three main types of engineering materials: metals, ceramics and polymers. These materials can we combined to form a composite(1). This takes advantage of the different properties of each material to develop a composite material with better overall performance than the original materials. The effects of thermal treatment on the mechanical properties of matrix–jute composites showed that laminates treated at 60°C had better mechanical properties than those treated at room temperature or 100°C. The finite element method (FEM) predictions matched the experimental data very closely under both stress and strain conditions(2). Jute, one of the most extensively studied natural fibers employed as reinforcement in polymer composites, is a by-product of the Corchorus capsularis plant and contains approximately 60% cellulose, 22% hemicellulose, and 16% lignin(3). For test and FEM ,they used in Abaqus to examine the mechanical properties of eco-friendly epoxy composites reinforced with jute steel fillers and fibers. Toughness of the composites relies on different ratios of fillers, particularly in ballistic applications. Jute reinforced composites have greater durability than to those filled with steel(4).Hybrid composites has enhanced performance tensile, flexural strength, also impact resistance which is important for various applications(5). These composites enhance mechanical properties across various engineering applications(6). Combining diffrent fiber like jute, with glass fibers is a promising approach for sustainability. Performance of these composites depends strongly on fiber orientation(7). Certain properties of jute fiber reinforced composites, such as impact resistance, remain unexplored despite their critical importance in applications like automotive crashworthiness (8). ASTM D638 provides standardized methods for measuring tensile strength, elongation, and other properties , which are essential for quality assurance and material selection across various applications (9). Hybrid composites made of glass fibers which enhance the mechanical performance and jute fibers which improve ductility, while tire fibers contribute to toughness and environmental benefits. These composites show great potential for light weight, high performance applications(10).

Materials and Methods

Design and manufacturing. The contact molding method was chosen for its advantages, including wide shape possibilities, no dimensional limitations, a smooth, gel coated surface, and moderate to good mechanical properties.

Presentation of materials. The developed composite material is based on an epoxy resin matrix reinforced with fibers (jute, tire, glass) (Fig.1).

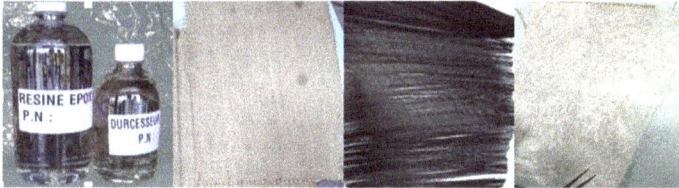

Fig. 1 The matrix epoxy resins with reinforcement using fibers of jute, tire, and glass.

Preparation of composite material. The fibers are manually cut into several sections according to the module's proportions, and the epoxy matrix is prepared. To ensure proper curing, 50% epoxy resin is mixed with 50% hardener. A total of 180g is prepared by combining 90g of epoxy resin with 90g of hardener (Fig.2).

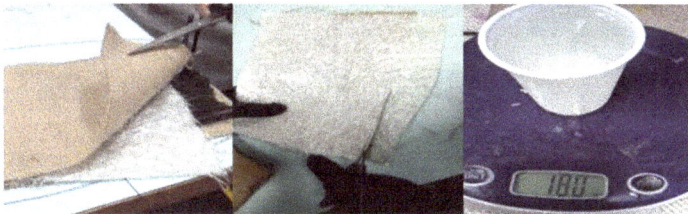

Fig. 2 JTG material.

To achieve a better surface finish and help extract the composite, a sheet of Mylar is placed under the mold, as well as assist with impregnating the fibers evenly with the epoxy matrix. A deairing roller is used to spread the epoxy mix, followed by adding tire fibers on top of the previous layer along with glass fibers. Once completed, the samples are placed under concrete bars to help get out any air bubbles that may be in the fibers (Fig.3).

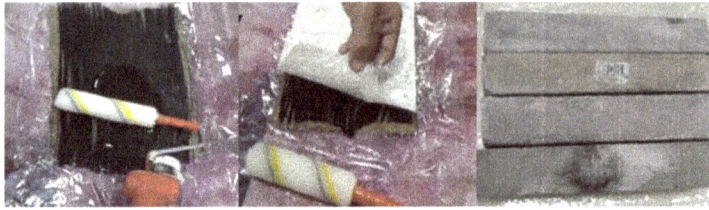

Fig. 3 The Sheet of Mylar.

Experimental tests

Laser cutting method. Laser cutting has the advantages of a clean and precise cut with little waste of material, makes it an excellent option for shaping fiber reinforced composites. It provides a high level of repeatability and minimizes post processing. It maximizes the efficiency while reducing heat damages (Fig.4).

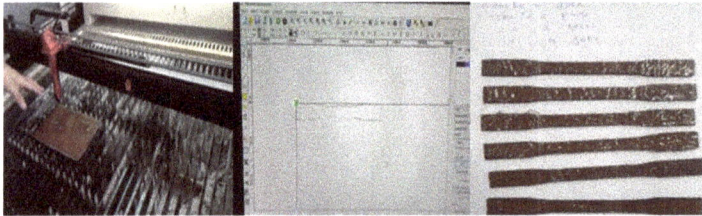

Fig. 4 Cutting laser..

Hardness . The Shore hardness method is utilized for determining the resistance of materials and elastic deformation. It is also used to assess the mechanical properties of silicone rubber during production (11). Durometer tests are calibrated for Shore A (blue = 31, yellow = 61.3, gray = 91.2) and Shore D (blue = 22.3, gray = 39, and black = 82.6) to check for accuracy. The hardness is measured on a composite specimen on both faces in two separate locations. The penetration depth is noted and converted to a Shore hardness value. The Shore hardness value falls within a range of 0-100, which is indicated on the durometer (Fig. 5).

Fig. 5 Hardness test.

Optical microscopy. Microscopy allows to observe deformations and damage mechanism in these materials, which provides useful knowledge for modeling and simulating the manufacturing process (12).

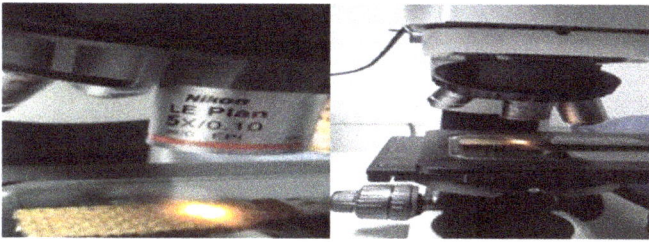

Fig. 6 The optical microscopy test.

Tensile test. This test is important for determining the tensile strength, also seeing the modulus of elasticity and failure modes of different composite materials, that are fastly grow in engineering applications, because of their strength to weight advantage and adaptability, based on the standardized(13).

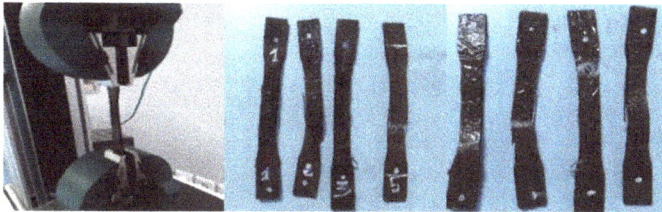

Fig. 7 Tensile test.

Simulation

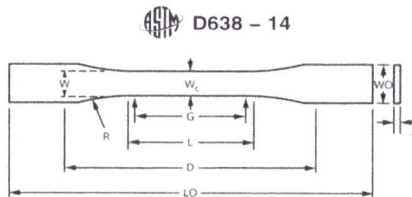

ASTM D638 – 14

Fig. 8 The studied composite material.

Abaqus was used to simulate the composite material by integrating experimental data, including mechanical properties. The 3D model was imported, and the fiber thickness of the composite structure was set to 0.1 cm, maintaining the same orientation as in the experimental test. Additionally, the boundary conditions were applied, with a fixed constraint at one end and a load applied at the opposite extreme. The simulation results served to assess deformations, constraints, and material behavior under applied forces (Fig 8 and 9).

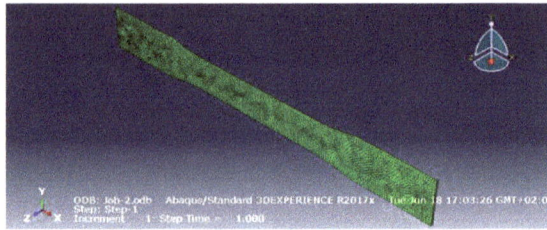

Fig. 9 The simulated composite material.

The optimization aims to maximize Young's modulus, mechanical strength, and initial resistance while minimizing elongation to ensure improved material performance. This structured approach ensures the development of a composite with enhanced mechanical properties, suitable for various engineering applications.

Results

Results of the experimental test. We use the rule of mixtures to calculate the volume fractions of fibers and matrix (Table.1).

Table 1.Material properties.

	ρ [g/cm3]	W[g]	V[cm3]	Volume Fration	E [GPA]	Poison's ratio	Transverse
Matrix epoxy	1,2	180	150	0.755	3.76	0.35	0.38
Jute	1,3	30	23.08	0.116	30	0.3	0.4
Tire	0,9	15	16.67	0.084	0.2	0.45	0.5
Glass	2,5	22	8.8	0.045	70	0.22	0.25

Results of hardness test. The Shore hardness test is a widely used method for evaluating the hardness, which uses Shore A and Shore D scales for distinct material types. The average hardness of shore-A value is 74.58, whereas the average of hardness shore D value is 39.54.

Results of the microscopy test. The first number (10 and 5) represents the zoom and how much time and the second number (0.25 and 0.1) reveals the optical properties of the microscope lens (Fig.10).

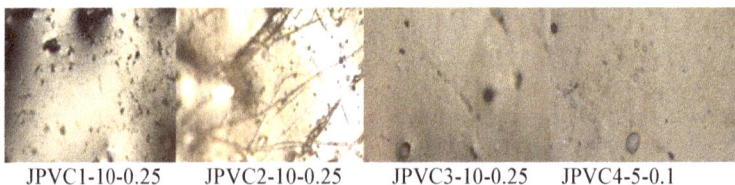

JPVC1-10-0.25   JPVC2-10-0.25   JPVC3-10-0.25   JPVC4-5-0.1

Fig. 10 The microscopy test.

Emerging Research in Materials for Environment, and Civil Infrastructure - GeoME 5.5   Materials Research Forum LLC
Materials Research Proceedings 58 (2026) 70-77   https://doi.org/10.21741/9781644903933-10

Results of tensile test

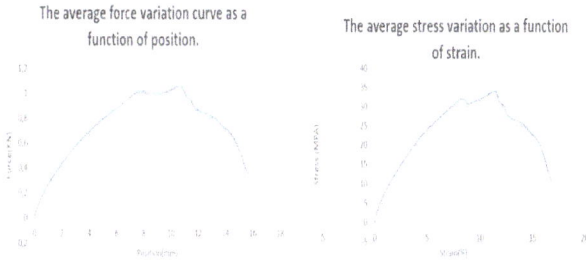

Fig. 11 The tensile test.

-Up to 1 kN (elastic deformation), the force increases gradually on the curve before plateauing steadily (load resistance). Cracks start to show up at 12 mm, which causes the force to drop until it fails at 16 mm. After reaching a load limit, the composite progressively deforms and breaks, despite its initial strong resistance.

- A stable plastic phase occurs at about 35 MPa after the composite experiences elastic deformation up to 32 MPa. Cracks start to show after 12%, and the stress gradually drops until it reaches 15%, at which point it completely fails.

Results of simulation with ABAQUS

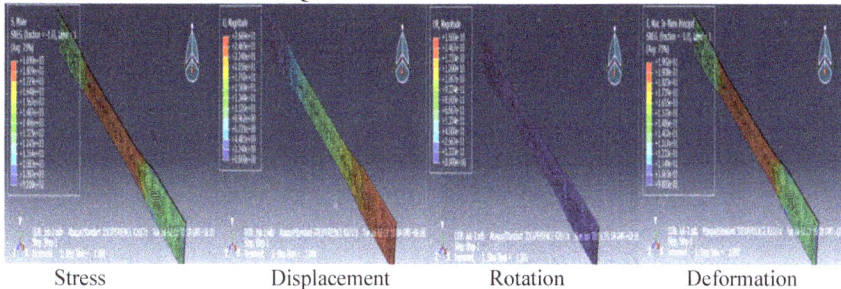

| Stress | Displacement | Rotation | Deformation |

Fig. 12 The results of the simulation.

Stress: The range of stresses is 1890 MPa (red) to 921.8 MPa (blue). The areas where the force is applied and close to the fixed end have the highest concentrations of stress. 75% of the maximum value is the average stress.

Displacement: The range of displacements is 2.689e-5 m (red) to 0 m (blue). The end, where the concentrated force is exerted, experiences the most displacement. The displacement is around 1.344e-5 m on average.

Rotation: The range of rotations is 1.600e-4 rad (red) to 0 rad (blue). The end, where the concentrated force is exerted, rotates the most. It rotates around 8.000e-5 rad on average.

Deformation: The range of maximum deformations is 9.803e-02 (blue) to 1.992e-01 (red). The fixed end and the force application point are where the big-gest deformations occur. The deformations average 75% of the maximal value (Fig.12).

Discussion

The study evaluates the mechanical properties of a JPVC composite, highlighting its durability, hardness, and structural integrity. With 75.5% epoxy content, the composite maintains strength while gaining flexibility from jute and tire fibers, and a small 4.5% glass fiber content further reinforces it. Moderate hardness values (Shore A: 74.58 HCR, Shore D: 39.54 HCR) suggest a balance between stiffness and flexibility, with jute and tire fibers improving impact resistance. Microscopy analysis confirms a perpendicular fiber alignment, good adhesion, and minor defects like pores and microcracks. Tensile tests indicate elastic de-formation up to 1 kN, followed by plasticity at 35 MPa, before fracturing at 16 mm elongation, with a ductile to brittle transition around 12% elongation.

Numerical simulations using Abaqus confirm high resistance to concentrated loads, with moderate displacements and rotations, ensuring structural rigidity and proving its suitability for industrial applications.

Conclusion

This study developed and optimized the JPVC composite, composed of an epoxy matrix reinforced with jute, tire, and glass fibers. A bidirectional fiber arrangement was confirmed through microscopic examination, which also revealed defects such as voids, air bubbles, and microcracks, indicating potential areas for manufacturing improvements. Mechanical tests demonstrated well balanced properties, with Shore A and D hardness values of 74.58 and 39.54 respectively. Tensile testing highlighted variations influenced by fiber distribution and matrix inconsistencies. Numerical simulations using Abaqus showed optimal stiffness up to a critical temperature, beyond which degradation affected performance, this study confirmed the potential of JPVC for lightweight, durable, and environmentally friendly applications. Future research will focus on improving fiber distribution, optimizing material refinement, and evaluating long term durability under cyclic loading and environmental conditions.

Acknowledgments

-This research is supported by the German International Cooperation: (GIZ) -RELANCE VERTE 2025".
- This research is supported by PhD-ASsociate Scholarship – PASS.

References

[1] Guggilla P, K, Novel Electroceramic: Polymer Composites - Preparation, Properties and Applications, in: Cuppoletti J (Ed.), Nanocomposites and Polymers with Analytical Methods [Internet], InTech, 2011. https://doi.org/10.5772/17049

[2] Ouchte I, Chafiq J, Ait El Fqih M, Chakir H, Effect of thermal treatment on mechanical properties and thermogravimetric analysis of laminate composite jute/epoxy, Mater Today Proc. 66 (2022) 135–139. https://doi.org/10.1016/j.matpr.2022.04.209

[3] Velu S, Joseph JK, Sivakumar M, Bupesh Raja VK, Palanikumar K, Lenin N, Experimental investigation on the mechanical properties of carbon-glass-jute fiber reinforced epoxy hybrid composites, Mater Today Proc. 46 (2021) 3566–3571. https://doi.org/10.1016/j.matpr.2021.01.333

[4] Zaini M, Ait El Fqih M, Mohamed MI, Bouab W, Chafiq J, Experiment and Simulation Study of the Effect of Fillers on Mechanical Properties of Sustainable Composite, Strojnícky Časopis - J Mech Eng. 72(2) (2022) 239–246. https://doi.org/10.2478/scjme-2022-0032

[5] Hasan MM, Islam MA, Hassan T, Analysis of jute-glass fiber reinforced epoxy hybrid composite, Heliyon. 10(24) (2024) e40924. https://doi.org/10.1016/j.heliyon.2024.e40924

Emerging Research in Materials for Environment, and Civil Infrastructure - GeoME 5.5   Materials Research Forum LLC
Materials Research Proceedings 58 (2026) 70-77   https://doi.org/10.21741/9781644903933-10

[6] Patil ABV, Estimation of Tensile and Flexure Strength of Hybrid Composites along with Epoxy as Matrix and Jute & E-Glass as Reinforcements, Int J Innov Sci Res Technol IJISRT. (2024) 2528–2533. https://doi.org/10.38124/ijisrt/IJISRT24OCT1836

[7] Rahman A, Hasib MdA, Islam MdA, Alam I, Chanda S, Fabrication and Performance Investigation of Natural-Glass Fiber Hybrid Laminated Composites at Different Stacking Orientations, J Nat Fibers. 20(1) (2023) 2143981.https://doi.org/10.1080/15440478.2022.2143981

[8] Karthi N, Kumaresan K, Sathish S, Prabhu L, Gokulkumar S, Balaji D, et al., Effect of weight fraction on the mechanical properties of flax and jute fibers reinforced epoxy hybrid composites, Mater Today Proc. 45 (2021) 8006–8010.https://doi.org/10.1016/j.matpr.2020.12.1060

[9] Tatus NA, Polilov AN, Vlasov DD, Akhmedshin EKh, A special specimen shape for adequate determination of tensile strength of a unidirectional fiber-reinforced composite, Ekaterinburg, Russia (2020) 040044. https://doi.org/10.1063/5.0036628

[10] Kavvuru C, Devakumar MLS, Mechanical Property Evolution of Jute and E-glass fiber Hybrid Polymer Matrix composites, Int J Sci Res Sci Eng Technol. (2019) 115–121.https://doi.org/10.32628/IJSRSET196230

[11] Zhao H, Allanson D, Ren XJ, Use of Shore Hardness Tests for In-Process Properties Estimation/Monitoring of Silicone Rubbers, J Mater Sci Chem Eng. 03(07) (2015) 142–147.http://dx.doi.org/10.4236/msce.2015.37019

[12] Krasnobrizha A, Rozycki P, Cosson P, Gornet L, Modélisation des mécanismes d'hystérésis des composites tissés à l'aide d'un modèle collaboratif élasto-plastique endommageable à dérivées fractionnaires, in: Godin N, Boisse P, editors, Matér Tech. 104(4) (2016) 407.https://doi.org/10.1051/mattech/2016018

[13] Geng J, Lyu J, Cai Y, Thermomechanical Properties of Ramie Fiber/Degradable Epoxy Resin Composites and Their Performance on Cylinder Inner Lining, Materials. 17(19) (2024) 4802. https://doi.org/10.3390/ma17194802

Emerging Research in Materials for Environment, and Civil Infrastructure - GeoME 5.5   Materials Research Forum LLC
Materials Research Proceedings 58 (2026) 78-85   https://doi.org/10.21741/9781644903933-11

Comparative study of the properties of mortar incorporating sugarcane bagasse ash and natural pozzolan

Fatima BOUKABOUS[1a], Jalal IKEN[1b], Omar DADAH[1c], Khalil NACIRI[1d], Issam AALIL[1e], Ali CHAABA[1f]

[1]University of Moulay Ismail, National Higher School of Engineering (ENSAM), Meknes, Morocco

[a]f.boukabous@edu.umi.ac.ma, [b]j.iken@edu.umi.ac.ma, [c]omardadah8@gmail.com, [d]khalil.nac@gmail.com, [e]aalil.issam@gmail.com, [f]m.chaaba@ensam.umi.ac.ma

Keywords: Pozzolanic Materials, Sugarcane Bagasse Ash, Natural Pozzolan, Mortar, Cement Replacement, Porosity, Compressive Strength

Abstract The cement industry is facing numerous environmental and energy-related challenges. Its fabrication requires a significant amount of energy and causes various environmental impacts, including high consumption of natural resources, the release of dust and certain pollutants, as well as significant greenhouse gas emissions. Nearly 50% of global CO_2 emissions are attributed to cement plants, and approximately 900 kg of CO_2 are released per ton of cement generated. While a complete replacement of cement is difficult, partial substitution is possible through Utilizing pozzolanic substances. These materials can improve resistance and durability of mortar. The purpose of this study is to valorize two pozzolanic materials: sugarcane bagasse ash (SCBA) and natural pozzolan (PN). The research involves comparing the structural and physical properties of mortars formulated with these two materials. Mortars were prepared with cement replacement rates of 0% and 10% for each material. The results showed that variations in physical properties remained small. The formulation with 10% SCBA provided the best overall balance: it combined high density, good ultrasonic pulse velocity, satisfactory compressive strength, and only a minor reduction in strength in flexion when compared to the control formula. Meanwhile, the inclusion of 10% PN improved mortar porosity, ultrasonic velocity, and compressive strength, although its effect on flexural strength was less significant. This study demonstrates that using pozzolanic materials such as bagasse ash and natural pozzolan can reduce cement consumption while maintaining good mortar performance, thereby contributing to more sustainable and environmentally friendly construction practices.

1. Introduction

Cement is extensively used as a primary binder in the construction industry due to its availability and performance. It is used as the primary binder in construction industries across the globe. The global cement production in 2023 was estimated at approximately 4,100 million tons [1]. However, it should be noted that this industry causes major environmental problems, namely high energy consumption and the emission of a significant percentage of greenhouse gases (GHG). Cement industry cause more 8% of (GHG) emissions. [2]. In addition, the cement sector is considered the third international energy consumption sector after steel and aluminum[3]. Faced with these environmental constraints, Multiple research has been conducted to develop solutions that can reduce the ecological impact of the cement plant on the environment, in particular the partial substitution of cement with pozzolanic materials.

The main components of pozzolanic materials are siliceous or silico-aluminous substances. These materials can exhibit characteristics similar to cementitious materials when they react chemically with calcium hydroxide under ambient conditions, especially if they are finely ground. [4]. Pozzolans are classified into two main categories: natural pozzolans of volcanic or sedimentary origin [5], and agricultural pozzolans, namely SCBA (silica-rich agricultural residue), which forms a favorable replacement to cement in the production of concrete or mortar, while improving their mechanical and physical properties and subsequently contributing to the sustainable management of agricultural waste [6].

Berenguer , Kasaniya et al.[7, 8], conducted studies on SCBA and PN to determine their potential as cement additives. They found that both materials are rich in silica and or alumina and can reduce cement usage in mortar formulations while improving certain properties of the final material.

In order to evaluate these two materials, a comparative study of their mechanical performance, particularly their compressive and flexural strength, as well as their physical characteristics (particularly density and porosity), was carried out on mortars incorporating these two pozzolanic materials. The experimental methodology consists of preparing mortars in which the cement has been partially replaced by CBA and natural pozzolan at a rate of 10%, while maintaining a reference mixture without substitution (0%).

2.   Materials and Methods

2.1   Materials

2.1.1   Sugarcane bagasse ash

SCBA is a material obtained after the complete combustion of sugarcane bagasse, which is a biomass residue resulting from the mechanical extraction of juice from sugarcane. The SCBA employed in this present investigation was collected from an industrial sugar refinery situated in the LGHRAB region. After collection, it was air-dried and then sieved to remove coarse particles. An additional grinding process was carried out using a mill to increase its fineness and enhance its pozzolanic reactivity (Fig.1).

Fig.1.The ash after combustion

In this study, XRD analysis was used to evaluate, in qualitative terms the crystalline phases present in the SCBA. This ash exhibits crystalline phases of quartz and cristobalite as shown in (Fig. 2)

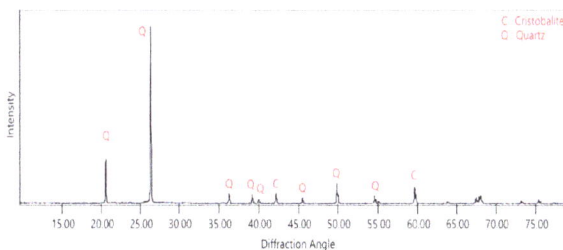

Fig. 2.XRD patterns of SCBA

2.1.2 Natural Pozzolan

Natural pozzolan (NP) is a volcanic rock of magmatic origin, formed primarily by the rapid cooling of lava upon exposure to water or air. It is generally red, black, or gray in color and has a porous, lightweight structure. Its name comes from the city of Pozzuoli in Italy, where it was widely used during the Roman era to produce durable hydraulic concrete. In this Research we are used a natural pozzolan which was sourced from the Midelt Province, known for its volcanic outcrops and characteristic black color. It was ground in the laboratory and then sieved (Fig.3).

Fig.3. Natural pozzolan from Midelt

2.1.3 Cement

The cement CPJ 45 used in this research supplied by Lafarge-Holcim. It consists of at least 65% clinker, complemented by additives such as limestone or fly ash. According to the technical datasheets from the factory [9], its minimum compressive strength after 28 days is 32.5 MPa , initial setting time higher 90 minutes, a dimensional stability not exceeding 10 mm, a density of 3 g/cm³, and an average particle size of 14 μm. Table [1] summarizes its chemical composition.

Table 1: chemical constituents of Cement

Component	Mass Percentage
Silica (SiO2)	20-25%
Alumina (Al2O3)	4-6%
Iron Oxide (Fe2O3)	2-5%
Calcium (CaO)	60-67%
Magnesium (MgO)	0.5-2.5%
Sulfur(SO3)	1-3%
Anhydrite (CaSO₄)	1-3%
Alkalis (Na2O, K2O)	0.5-1.5%

2.1.4 Sand

The sand utilized is a reference siliceous sand, in accordance with the NF EN 196-1 standard [10], characterized granulometry ranging from.0.08 mm to 2 mm table [2]. It is sourced from the Addarouch quarry in the El Hajeb region.

Table 2: Granulometric Composition of the Sand Used

Sieve Mesh Size (mm)	2	1.25	0.8	0.5	0.16	0.08
Cumulative Retained (%)	0	46	65	78	95	100

2.1.5 Water

The water used for mixing is potable water.

Emerging Research in Materials for Environment, and Civil Infrastructure - GeoME 5.5   Materials Research Forum LLC
Materials Research Proceedings 58 (2026) 78-85   https://doi.org/10.21741/9781644903933-11

2.2   Mix Design

Three mortar formulations were prepared: FT (Control): 100% cement, FSCBA: 90% cement + 10% SCBA, FPN: 90% cement + 10% NP. The ratio of binder to sand was set at 1/3, following the NF EN 196-1 [10] standard, and the ratio water of binder was adjusted in order to obtain a flow spread of 145 ± 5 mm, measured using the table used for flow testing. For each formulation, six samples measuring 4 × 4 × 16 cm were cast. The samples were demolded after 24 hours and then followed by a water cure at 20 ± 2°C till the age of the teste. Characterization was performed at 28 days. Table [3] presents the adopted mix proportions.

Table 3: Mix design of the mortar

Type	FT	FSCBA	FPN
cement (g)	1100	990	990
SCBA(g)	0	110	0
PN(g)	0	0	110
Water(g)	550	550	550
sand(g)	3300	3300	3300

2.3 Test procedures

2.3.1 Porosity and bulk density

To determine the porosity and bulk density after 28 days, the test samples were surface-dried with a damp towel to eliminate excess water, then weighed to determine the saturated (Ms). Next, the mass Mhyd was determined by means of hydrostatic weighing. Then the samples were dried in an oven at 100°C ± 1 °C till the weight remained constant (variation less than 0.1% over 48 hours). This final mass is referred to as Mdry (dry mass)

The porosity is given by:

$$P = \frac{M_{sat} - M_{dry}}{M_{sat} - M_{hyd}} \times 100 \quad (1)$$

The bulk density ρ is determined by:

$$\rho = \frac{M_{dry}}{M_{sat} - M_{hyd}} \quad (2)$$

2.3.2   Compression and flexion strengths

For each sample, compression and flexion strengths were determined in compliance with the NF EN 196-1 standard [10] , which specifies the Characterization methods for hydraulic cements. The test was performed using a Proeti electromechanical press, At a loading rate of 0.5 mm/min.

The three samples' flexural strength Was determined using the following equation:

$$R_f = \frac{1.5 \times F_f \times l}{b^3} \quad (3)$$

Where Rf   is the failure load of the specimen in bending, L= 100mm, and b=40mm.

Compressive strength tests were performed on half-prisms following the NF EN 196-1 standard [10].The compressive strength is determined as follows:

$$R_c = \frac{F_c}{b^2} \quad (4)$$

Where Rc is the breaking load of the specimen in compression and b= 40 mm.

The three remaining half-prisms were used to determine porosity, bulk density, and then compressive strength in the dry state.

Emerging Research in Materials for Environment, and Civil Infrastructure - GeoME 5.5   Materials Research Forum LLC
Materials Research Proceedings 58 (2026) 78-85   https://doi.org/10.21741/9781644903933-11

3.   Results and discussion

3.1   Apparent density and Porosity

The porosity and densities among the three formulations are shown in (Fig. 4)

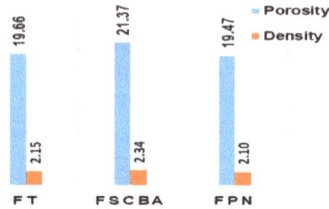

Fig. 4. Porosity (%) and Density

For the FSCBA formulation, an increase in porosity is observed relative to the control formulation. This may be due to the fibrous texture or cellular structure of SCBA particles, which are often irregular and poorly reactive if not adequately treated (finely calcined) [11]. The increase in density may seem paradoxical, but it can be explained by the internal structure, which may contain fine particles capable of filling voids or air spaces within the mortar matrix, resulting in a denser material. This is known as the filling or packing effect [12].

As for the FPN formulation, the slight decrease in porosity compared to FT may be attributed to a filling effect, in which the finer natural pozzolan particles occupy the pores of the mix. Furthermore, the slight decrease in density despite the overall compactness caused by the the volcanic nature of pozzolan [13].

3.2   Compression and flexion strength.

According to (Fig.5) , we note an improvement in mortar's compression strength of 10% The SCBA, reaching 25.41 MPa, this exceeds the FT control formulation's value (21.05 MPa) and the FPN (22.44 MPa). The incorporation of 10% PN remains beneficial for compressive strength but is less effective than SCBA.

On the other hand, a decrease compared to FT was observed for flexural strength (11.40% for mortar containing 10% SCBA and 22.72% for mortar containing 10% NP).

We can therefore conclude that the incorporation of SCBA resulted in a minor decrease in flexural strength, but a significant progress in compressive strength. This result indicates that SCBA exhibits pozzolanic activity that promotes the formation of secondary cementitious reaction products, thereby densifying the matrix. In contrast, mortar containing natural pozzolan (FPN) exhibits intermediate values for compression strength and a significant reduce in flexion strength

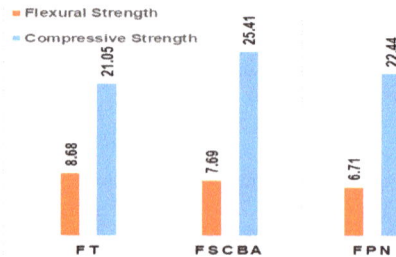

Fig. 5. Compressive and Flexural Strength (MPa)

Emerging Research in Materials for Environment, and Civil Infrastructure - GeoME 5.5   Materials Research Forum LLC
Materials Research Proceedings 58 (2026) 78-85   https://doi.org/10.21741/9781644903933-11

The dissipation of energy under a bending load due to the non-uniform distribution of the interfacial transition zones (ITZ) between the cement paste and the aggregates and the heterogeneity of the microstructure can lead to a reduction in the flexural strength recorded in mortars incorporating the two pozzolanic additives between the cement paste and the aggregates (Chen andt al. 2024). However, The material's capacity to transmit tensile forces may be limited by a weakening of the link among the pozzolanic particles and the cement matrix, thereby facilitating the initiation and propagation of cracks [15]. Furthermore, an excess of non-reactive or weakly reactive particles may introduce weak zones, which compromise the overall cohesion of the matrix and negatively affect its flexural mechanical performance[16].

Several research has explored to analyze the effet of incorporating (SCBA) on the compressive performance of mortars and concretes. The studies aim to assess how the use of SCBA as a cement substitute affects Strengths of materials, particularly in terms of strength gain or loss over the short and long term. The obtained values fluctuate with the replacement rate, the particle size of the ash, as well as the curing conditions and the mix design [17,18]. Muhammad Nasir Amin el al. [19] demonstrated in studies on the development of fiber-reinforced cement composites, that 10% cement replacement with ground SCBA provides the greatest resistance to compression. K. Ganesan el al. [20] examined the issue of waste disposal from agro-industrial sources, namely, wheat straw ash, rice husk ash, and sugarcane bagasse ash. This research was centered primarily on the use of SCBA as a binder material, and the results showed that The optimal compressive strength was observed when 10% of the cement was substituted with (SCBA).As reported by Senhadji el al [21], at an early age, natural pozzolan (NP) tends to reduce strength. However, at later ages, the pozzolanic reaction contributes to strength development that surpasses that of ordinary Portland cement, particularly after 28 days [22, 23]

3.3   Pozzolanicity activity index (PAI)

The pozzolanic activity index (PAI) for compression is given by the formula below, where $R_C(Fi)$ and $R_C(F0)$ are the compression strengths of mortar with the addition of pozzolan and the reference mortar, respectively:

$$PAI\ (Fi) = \frac{R_C(Fi)}{R_C(F0)} \times 100 \quad (5)$$

A similar index could be defined using flexural strength instead of compressive strength.

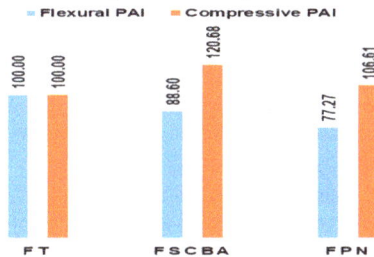

Fig. 6. Compressive and Flexural PAI (%)

According to (Fig.6), The incorporation of SCBA at a 10% substitution rate improves the compressive strength activity index (PAI) in comparison with the control (FT), reaching a value of 120.68%, while the flexural strength index decreased to 88.60%. These results indicate that SCBA exhibits notable pozzolanic activity at a 10% replacement level, in accordance with ASTM C618-19 [24]. A similar trend is observed with the 10% replacement by natural pozzolan, with registering compressive and flexural resistance indices of 88.60% and 77.27%, respectively. However, these results are less favorable than those obtained with the 10% SCBA addition.

Conclusion

1. The results indicate that variations in physical properties remain relatively small between the different formulations.
2. The mortar incorporating 10% sugarcane bagasse ash (SCBA) offers the best overall compromise. It showed a significant improvement in density and compressive strength and a moderate decrease in flexural strength in contrast to the reference mortar.
3. The incorporation of natural pozzolan (NP) into mortar positively affects the mortar's porosity and compressive strength. However, its impact on flexural strength remains slightly negative.
4. Using SCBA and natural pozzolan, as cement particle substitutes, reduces the environmental impact associated with the cement sector. This approach fulfills one of the principles of sustainable construction, which aims to balance the technical performance of resources with environmental protection

References

[1] « Mineral Commodity Summaries 2024 ».

[2] B. S. Thomas *et al.*, « Sugarcane bagasse ash as supplementary cementitious material in concrete – a review », *Mater. Today Sustain.*, vol. 15, p. 100086, nov. 2021. https://doi.org/10.1016/j.mtsust.2021.100086

[3] K. V. Teja, P. P. Sai, et T. Meena, « Investigation on the behaviour of ternary blended concrete with scba and sf », *IOP Conf. Ser. Mater. Sci. Eng.*, vol. 263, n° 3, p. 032012, nov. 2017. https://doi.org/10.1088/1757-899X/263/3/032012

[4] V. Furlan et Y. F. Houst, « Les matériaux pouzzolaniques et leur utilisation », *Chantiers Suisse*, vol. 11, n° 7, p. 29-32, 1980.

[5] M. Sharbaf, M. Najimi, et N. Ghafoori, « A comparative study of natural pozzolan and fly ash: Investigation on abrasion resistance and transport properties of self-consolidating concrete », *Constr. Build. Mater.*, vol. 346, p. 128330, 2022.

[6] S. E. L. Gudia *et al.*, « Sugarcane bagasse ash as a partial replacement for cement in paste and mortar formulation – A case in the Philippines », *J. Build. Eng.*, vol. 76, p. 107221, oct. 2023. https://doi.org/10.1016/j.jobe.2023.107221

[7] R. A. Berenguer, A. P. B. Capraro, M. H. F. de Medeiros, A. M. P. Carneiro, et R. A. De Oliveira, « Sugar cane bagasse ash as a partial substitute of Portland cement: Effect on mechanical properties and emission of carbon dioxide », *J. Environ. Chem. Eng.*, vol. 8, n° 2, p. 103655, avr. 2020. https://doi.org/10.1016/j.jece.2020.103655

[8] M. Kasaniya, M. D. A. Thomas, et E. G. Moffatt, « Pozzolanic reactivity of natural pozzolans, ground glasses and coal bottom ashes and implication of their incorporation on the chloride permeability of concrete », *Cem. Concr. Res.*, vol. 139, p. 106259, janv. 2021. https://doi.org/10.1016/j.cemconres.2020.106259

[9] « WePxh6LbXduKqJLXWJaW.pdf ». Consulté le: 12 mars 2025. [En ligne]. Disponible sur: https://www.cimat.ma/storage/projects/July2021/WePxh6LbXduKqJLXWJaW.pdf

[10] « NF EN 196-1 », Afnor EDITIONS.

[11] P. Jagadesh, A. Ramachandramurthy, R. Murugesan, et K. Sarayu, « Micro-analytical studies on sugar cane bagasse ash », *Sadhana*, vol. 40, n° 5, p. 1629-1638, août 2015. https://doi.org/10.1007/s12046-015-0390-6

[12] N. Chusilp, C. Jaturapitakkul, et K. Kiattikomol, « Utilization of bagasse ash as a pozzolanic material in concrete », *Constr. Build. Mater.*, vol. 23, n° 11, p. 3352-3358, nov. 2009. https://doi.org/10.1016/j.conbuildmat.2009.06.030

[13] A. Bellil, A. Aziz, I.-I. El Amrani El Hassani, M. Achab, A. El Haddar, et A. Benzaouak, « Producing of Lightweight Concrete from Two Varieties of Natural Pozzolan from the Middle Atlas (Morocco): Economic, Ecological, and Social Implications », *Silicon*, vol. 14, n° 8, p. 4237-4248, juin 2022. https://doi.org/10.1007/s12633-021-01155-8

[14] Q. Chen, J. Zhang, Z. Wang, T. Zhao, et Z. Wang, « A review of the interfacial transition zones in concrete: Identification, physical characteristics, and mechanical properties », *Eng. Fract. Mech.*, vol. 300, p. 109979, avr. 2024. https://doi.org/10.1016/j.engfracmech.2024.109979

[15] A. Joshaghani, A. A. Ramezanianpour, et H. Rostami, « Effect of incorporating Sugarcane Bagasse Ash (SCBA) in mortar to examine durability of sulfate attack ».

[16] P. Jagadesh, A. Ramachandramurthy, R. Murugesan, et T. Karthik Prabhu, « Adaptability of Sugar Cane Bagasse Ash in Mortar », *J. Inst. Eng. India Ser. A*, vol. 100, n° 2, p. 225-240, juin 2019. https://doi.org/10.1007/s40030-019-00359-x

[17] C. K. Gupta, A. K. Sachan, et R. Kumar, « Examination of Microstructure of Sugar Cane Bagasse Ash and Sugar Cane Bagasse Ash Blended Cement Mortar », *Sugar Tech*, vol. 23, n° 3, p. 651-660, juin 2021. https://doi.org/10.1007/s12355-020-00934-8

[18] C. K. Gupta, A. K. Sachan, et R. Kumar, « Utilization of sugarcane bagasse ash in mortar and concrete: A review », *Mater. Today Proc.*, vol. 65, p. 798-807, janv. 2022. https://doi.org/10.1016/j.matpr.2022.03.304

[19] M. N. Amin *et al.*, « Role of Sugarcane Bagasse Ash in Developing Sustainable Engineered Cementitious Composites », *Front. Mater.*, vol. 7, avr. 2020. https://doi.org/10.3389/fmats.2020.00065

[20] K. Ganesan, K. Rajagopal, et K. Thangavel, « Evaluation of bagasse ash as supplementary cementitious material », *Cem. Concr. Compos.*, vol. 29, n° 6, p. 515-524, juill. 2007. https://doi.org/10.1016/j.cemconcomp.2007.03.001

[21] Y. Senhadji, Escadeillas ,Gilles, Khelafi ,Hamid, Mouli ,Mohamed, et A. S. and Benosman, « Evaluation of natural pozzolan for use as supplementary cementitious material », *Eur. J. Environ. Civ. Eng.*, vol. 16, n° 1, p. 77-96, janv. 2012. https://doi.org/10.1080/19648189.2012.667692

[22] F. P. Alishah et M. M. Razaei, « Effect of Natural Pozzolan on Concrete's Mechanical Properties and Permeability in Various Grades of Cement ».

[23] Z. Makhloufi, M. Chettih, M. Bederina, E. l. H. Kadri, et M. Bouhicha, « Effect of quaternary cementitious systems containing limestone, blast furnace slag and natural pozzolan on mechanical behavior of limestone mortars », *Constr. Build. Mater.*, vol. 95, p. 647-657, oct. 2015. https://doi.org/10.1016/j.conbuildmat.2015.07.050

[24] « Standard Specification for Coal Fly Ash and Raw or Calcined Natural Pozzolan for Use in Concrete ».

Emerging Research in Materials for Environment, and Civil Infrastructure - GeoME 5.5   Materials Research Forum LLC
Materials Research Proceedings 58 (2026) 86-91   https://doi.org/10.21741/9781644903933-12

The use of timber to enhance seismic performance in vernacular architecture of Morocco: The case of the Haouz Valley

Assia LAMZAH[1,a *], Ismail MAKHAD[2,b]

[1]National School of Architecture, Rabat, Morocco

[2]University of Ottawa, Civil Engineering Department, Ottawa, Canada

[a]a.lamzah@enarabat.ac.ma, [b]imakh034@uottawa.ca

Keywords: Vernacular, Architecture, Seismic Performance, Earth, Timber, Haouz Valley, Morocco

Abstract. Vernacular Architecture is the product of the geographical, social, economic, and cultural contexts. It is mostly presented as eco-friendly, functional, and community-driven. Additionally, it is constructed with empirical knowledge accumulated by building masters using local resources, which may bear important lessons. Morocco has a rich vernacular architecture that witnesses its history and local identity. It testifies of a traditional design wisdom, great capacity of adaptability to the context, and a great genius and know-how that developed over centuries in different regions and territories of the country. One of the interesting aspects of these traditional construction methods and design wisdom is the use of local materials to enhance the seismic performance of the buildings. More precisely, timber was used in earthen and stone architecture to reinforce the structure and enhance its seismic performance. After the Haouz earthquake, in September 2023, such a technique was the reason for the preservation of several constructions. The present paper focuses on the use of timber in traditional construction techniques of the valley, as an old and effective technique to enhance the seismic performance of the structure. More precisely, it presents and analyses specific techniques, mainly in the walls, both is corners and, in the openings, (doors and windows where timber elements are fixed to the masonry walls using mechanical point-to-point connections). The objective is to not only present these techniques as construction and architectural vernacular heritage, but also to analyze the possibility to (re) using them in the ongoing reconstruction projects in a more sustainable approach.

Introduction

Vernacular architecture embodies centuries of empirical knowledge adapted to local environmental and social contexts, often resulting in sustainable and functional buildings [1]. In seismically active regions, such as Morocco's Haouz Valley, traditional construction methods reveal an inherent capacity to withstand earthquakes through the strategic use of timber elements embedded within predominantly earthen and stone structures. The 2023 Haouz earthquake highlighted the structural deficiencies of many modern and contemporary buildings but also emphasized the resilience of vernacular constructions that utilize timber reinforcements. This paper examines these traditional timber applications and advocates for their revival in contemporary seismic-resistant architecture, to be used in the Haouz valley for a more resilient architecture.

Contextual Background

The Haouz Valley, located at the foothills of the High Atlas Mountains, is a tectonically active region prone to recurring seismic events. The valley's settlements mostly present rural villages in which the constructions are typically built in pisé, adobe, and stone. Generally speaking, these materials are identified as being thermic and phonic efficient but offer limited seismic ductility. Northern Morocco showcases several vernacular architectures; Haouz, like other Moroccan regions such as Rif, often exhibited load bearing earthen or stone walls with timber reinforcement. Roofs are flat or close-to-flat with timber beams locally named iskar'n . Timber as a material of construction possesses tensile strength and flexibility.

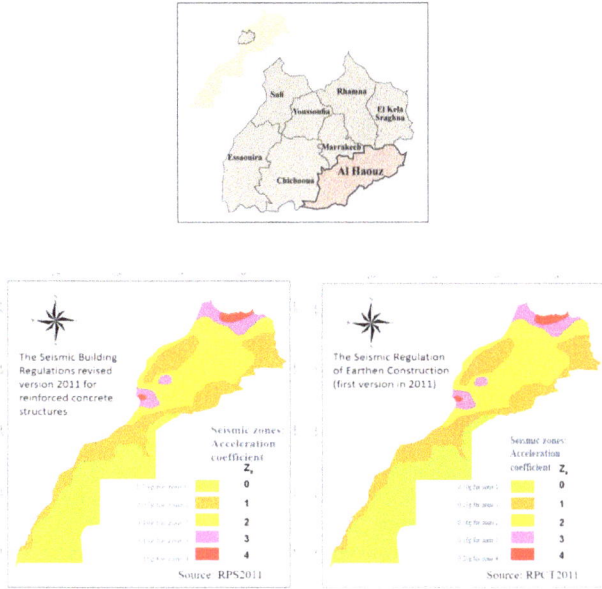

Fig. 1: Al Haouz situation and seismic zones in Morocco

These qualities, which timber holds, complement the compressive nature of earthen walls. During seismic events, these qualities offer better seismic vibrational behavior in terms of energy dissipation and loading redistribution. These techniques are, to us, interesting to explore further, from a technical and critical architecture standpoint, specifically considering the social and economic aspects of the community. The traditional builders produced vernacular architecture from local construction resources such as earth, stone, and straw. They created a form of place-specific architecture that guarantees sustainability and material-oriented physical stability. Timber was integrated inseparably into one of their material-oriented strategies for enhanced performance and stability, especially seismic safety. Timber use targeted the specific building zones of corners and openings. These techniques made the building connections flexible and contributed to mitigate destroying earthquake forces.

Emerging Research in Materials for Environment, and Civil Infrastructure - GeoME 5.5   Materials Research Forum LLC
Materials Research Proceedings 58 (2026) 86-91   https://doi.org/10.21741/9781644903933-12

Fig. 2: Example of timber reinforcement in walls in the Haouz

Seismic Performance of Vernacular Structures

Although earthen and stone walls are environmentally sound and architecturally characteristic, they may prove to be very fragile when under lateral seismic loads, which frequently lead to collapses. Architectural and landscape specificities are achieved by implementing timber reinforcements such as in facades. More importantly, because it boosts the building's ductility and controls crack developing, the building collapse is lessened out of control. As a result, using timber in vernacular architecture in the Haouz is not limited to its aesthetic and climatic qualities. This construction material serves as a structural element. Timber aids in the building's seismic resilience. This is because timber is strong and able to absorb energy. The material is thus useful for seismic strengthening in masonry structures. The effectiveness of timber for ductility has been examined in academic and experimental studies. For example, studies by the Getty Conservation Institute found that timber is utilized in historic buildings in Latin America to boost their seismic strength. They have also shown how these low-tech techniques allow timber to absorb and redistribute seismic forces. The material use of timber significantly reduces the risk of occurrence of big deformation and collapsing in masonry element. Additionally, they claim that the use of timber contributes to the conservation status of architectural heritage and landscape. For example, in the context of the vernacular architecture of Morocco, timber has been used in the wall and in the roof systems. Timber elements such as timber beams desiccate the building making it move as one piece. Timber can act as reinforcing agent at stress points including around corners or at orifice such as windows or doors; this practice has been supported by various empirical studies and structural analyses. Parisi and Piazza demonstrated how including timber element in a structure of masonry buildings enhances their seismic. Timber ductility allows the integration of the structure to control the deformation force resulting from seismic loads. Also, due to the compatibility with other materials including earthen and stone masonry, timber reduces the risk of delaminating or detachment between the materials. Also, practices vis-à-vis timber are aligned with the ancestral knowledge of Africans in southern Morocco. For instance, these techniques have been critical in seismic-prone areas and other continents such as Latin America.

Emerging Research in Materials for Environment, and Civil Infrastructure - GeoME 5.5   Materials Research Forum LLC
Materials Research Proceedings 58 (2026) 86-91   https://doi.org/10.21741/9781644903933-12

Fig. 3: Chaining and wooden sills of Almohad Monuments in southern Morocco [5].

The deployment of timber within the structural fabric of vernacular Moroccan buildings has a role of unifying and energy-absorbing element within masonry walls. Empirical Evidence from the 2023 earthquake revealed that buildings incorporating timber elements sustained less damage. Indeed, these buildings had less cracking and fewer collapses compared to purely masonry structures [3]. This empirical evidence validates the argument that integrating timber in earth structure, at specific points, has seismic benefits.

Vernacular Timber Techniques
In traditional masonry construction, particularly in seismic regions, one of the most critical points of vulnerability lies at the intersections of orthogonal walls. When timber beams are placed horizontally at wall intersections, they serve as ties that prevent separation. Therefore, this technique enhances structural coherence during seismic shaking. Indeed, under seismic loading, wall junctions tend to be separated, because of out-of-plane forces and differential movement between adjoining wall segments. Therefore, when we place horizontal timber beams at these intersections, it makes them act as structural ties, which enhances the coherence of the building envelope. In other words, these beams work as a mechanical linkage that distributes lateral forces across intersecting walls and reduces the risk of joint separation. This form of reinforcement not only increases the lateral stiffness of the building but also provides a degree of controlled flexibility, allowing the structure to dissipate seismic energy without building collapse.

First post 2023 earthquake damage assessments in the Houaz valley demonstrated that use of this technique contributed in maintaining the physical integrity of masonry structures during earthquake motion. Other earth buildings without this technique experienced severe damage. Even if the number of the buildings that used this technique and resisted the earthquake was not very important, this helped saving lives and architectural heritage of some historic buildings of some villages. Therefore, it represents a simple yet highly effective intervention. This technique needs to be further analyzed and reintroduced to contemporary construction principles, both in architectural heritage buildings preservation efforts and in contemporary vernacular construction, especially in rural vulnerable areas of the valley that suffered from the 20203 earthquake and are now under reconstruction.

Emerging Research in Materials for Environment, and Civil Infrastructure - GeoME 5.5   Materials Research Forum LLC
Materials Research Proceedings 58 (2026) 86-91   https://doi.org/10.21741/9781644903933-12

Fig. 4: Examples of timber framing around openings

Advanced technical evaluations showed that buildings incorporating timber framed doors and windows consistently demonstrate reduced damage levels and improved residual strength compared to those lacking such reinforcements. Moreover, this technique aligns with broader vernacular strategies that emphasize hybrid material systems, optimizing the complementary properties of timber and masonry to achieve seismic resilience.

Fig. 5: Example of Al Haouz building that resisted the 2023 earthquake

The strategic use of timber around doors and windows illustrates an interesting approach to seismic design. Traditional joint techniques fix timber elements into masonry at specific points, in order to control flexibility. In addition to the anti seismic performance, they reflect a cultural identity and reinforce landscape identity.

Implications for Contemporary Reconstruction

The post 2023 earthquake assessment showed that reintegrating timber-based seismic reinforcement in reconstruction presents several advantages on several aspects. On the one hand, it offers resilient architecture and contributes to save lives and to save the specificities of the architectural vernacular and heritage buildings. On the oher hand, it aligns with the economic vulnerabilities of such rural territories and provides an effective viable economic solution for the local population [2]. The contemporary reconstruction projects in the villages of the valley need to integrate these vernacular techniques. They also need to present them and inform local people of their effectiveness. Researchers need to think about how to preserve these techniques while answering local people's contemporary needs and right to development.

Reintroducing timber into seismic resilient construction strategies not only strengthens the physical resilience of reconstructed buildings but also supports cultural continuity and community engagement [1]. This vernacular technique demonstrates the integration of construction within broader ecological and cultural systems [2]. Moreover, this approach plays a critical role in preserving architectural heritage. These architectures and landscapes definitely contribute to preserve a community's sense of place and historical continuity.

Conclusion

In conclusion, the present research has analysed the use of timber, as local material, in the vernacular architecture of the Haouz valley, Morocco. The research's argument is that the integration of timber in the earthen architecture has the difference to make the architecture particularly more resilient. The Haouz earthquake of 2023 has pointed to many contemporary building's structural deficiencies and, on the contrary,. The vernacular buildings, which use timber reinforcements, have proven more resilient. Besides, the post-earthquake evaluation has shown that the research used guarantees many successful avenues in terms of structure, economy and culture. Therefore, in the conclusion, it is argued the necessity of integrating this vernacular knowledge in the contemporary practice of architecture and structural engineering to be possible to contribute to a more resilient architecture in the territory as vulnerable as the Haouz valley architectures.

References

[1] Oliver, P. (2006). Built to Meet the Needs: Cultural Issues in Vernacular Architecture. Routledge. https://doi.org/10.4324/9780080476308

[2] Aït Zamzami, et al. (2024). Resilience and Vulnerability: The Haouz Earthquake's Effect on Housing In The Western High Atlas of Morocco. International Journal for Disaster and Development Interface, 4(1), Pages 27–51. https://doi.org/10.53824/ijddi.v4i1.64

[3] Parisi, M. A., & Piazza, M. (2004). Seismic Strengthening of Traditional Timber Structures.13th World Conference on Earthquake Engineering, Vancouver, Canada.

[4] El Harrouni Khalid, Kharmich Hassane et Lamzah Assia. Seismic Performance of Traditional Urban Architecture in Morocco. In the International Journal of Heritage Architecture. 2017, Vol. 1, Issue 1, Pages 42-59. https://doi.org/10.2495/HA-V1-N1-42-59

[5] Basset H and H Terrasse. (1924). Sanctuaires et Forteresses Almohades. In Hesperis Tamuda, 4, Pages 9-91.

Emerging Research in Materials for Environment, and Civil Infrastructure - GeoME 5.5   Materials Research Forum LLC
Materials Research Proceedings 58 (2026) 92-98   https://doi.org/10.21741/9781644903933-13

Improving the non-destructive evaluation of in-situ concrete strength: The role of core location selection

Bouchra Kouddane[1,a] *, Youssef Jamil[2,b]

[1]Moroccan School of Engineering, Rabat, Morocco

[2]Higher School of Architecture of Casablanca, Casablanca, Morocco

[a]bouchrakouddane@gmail.com, [b]y.jamil@ecolearchicasa.com

Keywords: Core Sampling Strategies, Random Coring, In-Situ Concrete Strength, Concrete Strength Variability, Rebound Hammer Testing, Ultrasonic Pulse Velocity (UPV), Combined NDT Methods

Abstract. Assessing the strength of concrete in existing structures usually relies on a combination of non-destructive testing (NDT) and core-based destructive tests. This mixed strategy limits damage to the structure but still raises an important question for engineers: how can the measured in-situ strengths be considered statistically reliable? In this work, attention is directed to one crucial component of any investigation programme, namely the way in which core locations are selected. Several alternative sampling plans for core extraction are investigated, including conditional sampling, similarity-based sampling, proportional stratified sampling and a variance-oriented strategy. Their performance is analysed and compared with that of conventional random coring by means of an extensive statistical study. The methodology is applied to a large-scale case study on an existing building, where Rebound Hammer (RH) measurements and Ultrasonic Pulse Velocity (UPV) readings are collected and combined with core tests on structural members. The analysis focuses on two indicators of interest: the mean compressive strength and the associated variability. Two identification frameworks are considered: a bi-objective formulation when a single NDT method is used, and a multi-objective formulation when RH and UPV are jointly exploited. The various sampling plans are evaluated in terms of accuracy, efficiency and reliability, and systematically benchmarked against random coring. On this basis, the study identifies the strategies that yield the most robust estimates of in-situ concrete strength and proposes practical guidance to assist engineers in designing efficient and reliable investigation programmes for existing structures.

Introduction

Destructive testing is an experimental approach used to characterise the mechanical and physical properties of materials and structural elements by extracting cores from the member under investigation. It provides highly precise information on concrete strength and other properties, but at the cost of locally damaging the structure. Destructive tests do not only damage the structure locally; they also demand a lot of practical resources, from laboratory time to direct costs. For this reason, engineers try to limit the number of cores they take and increasingly rely on non-destructive testing (NDT) methods such as Ultrasonic Pulse Velocity (UPV) and Rebound Hammer (RH) [1,2]. In many case studies, several NDT techniques are used together in order to refine the estimation of in-situ concrete strength [1–3].

The quality of an investigation programme, however, is not determined by the test methods alone. It is strongly affected by the heterogeneity of the concrete in existing buildings, which makes it challenging to capture both the variability and the mean strength when only a limited number of measurements and cores can be obtained [4]. Uncertainties linked to in-situ measurements and their repeatability also need to be taken into account [5]. In addition, the

Emerging Research in Materials for Environment, and Civil Infrastructure - GeoME 5.5   Materials Research Forum LLC
Materials Research Proceedings 58 (2026) 92-98   https://doi.org/10.21741/9781644903933-13

locations chosen for core extraction can have a marked influence on the estimated strength distribution [6].

In practice, core locations are still often selected without making explicit use of the NDT measurement pattern [7]. Several authors have pointed out that this may introduce bias into the assessment of concrete strength [8,9]. If cores are restricted to a few accessible zones, the concrete in these areas may not reflect the overall strength level in the structure. Using NDT data in a more systematic way has therefore led to the concept of conditional coring, which aims to reduce such bias. For example, the study in [8] reports that conditional coring, where core locations are guided by NDT results, can reduce the number of required cores by up to three compared with random selection. Building on this idea, the present work goes one step further and investigates several alternative sampling plans for choosing core locations. By making a comparison of these sampling plans, the study aims to explain their inherent advantages and disadvantages in getting accurate estimates of concrete strength.

The evaluation methodology of sampling strategies is investigated using a population of size NT=100, derived from the experimental dataset presented by [9]. The statistical properties of the full dataset are taken as reference values and are assumed to represent the "true" behaviour of the material. The aim of the study is to reproduce a realistic investigation scenario: an engineer first performs non-destructive testing (NDT) at selected locations, then chooses a subset of these points for core extraction, measures the corresponding core compressive strengths, and finally calibrates a correlation model between NDT results and destructive test data. This model is subsequently used to predict compressive strength at locations where no cores are available. The predicted strengths are then compared with the reference values obtained from the full dataset, in order to evaluate the accuracy of the model.

Sampling plans of core locations
Conditional sampling:
Conditional coring relies on NDT results to guide the selection of core locations, so that the chosen cores cover as much as possible of the concrete strength range within the structure. All NDT measurements are first sorted in ascending order. For single-method configurations (RH only or UPV only), the 100 measurements are then ranked from 1 (lowest) to 100 (highest). In the combined approach, each point is assigned a composite rank obtained by averaging its RH and UPV ranks, as defined in Eq. (1) [10].

$$Rank(RH, UPV) = [Rank(RH) + Rank(UPV)]/2 \tag{1}$$

This procedure makes it possible to define specific groups of NDT measurements, referred to as NC groups, where NC denotes the number of cores used for model calibration. In practice, NC is selected in line with current engineering practice, typically between 3 and 20. After dividing the ranked dataset into NC groups of approximately equal size, one core location is randomly selected within each group.

Proportional stratified sampling:
In this sampling scheme, concrete strength is inferred from a population that is assumed to be stratified in space. The idea is that the structure can be divided into several strata, and that within each stratum the material properties are approximately independent and identically distributed. Under this assumption, the variability observed in each stratum is considered to be representative of that part of the population [11].

Operationally, the NDT dataset is first partitioned into a number of homogeneous groups according to a relevant characteristic, for example by grouping measurements into classes of high, medium and low NDT readings. Core locations are then chosen at random within each of these groups. The number of cores allocated to a given group is linked to its relative weight in the population, so that larger subgroups contribute more to the overall sample [12]. By decomposing a heterogeneous population into several more homogeneous subsets, stratified sampling can improve the precision of estimates for quantities of interest defined over the whole population [13].

Similarity sampling:
The purpose of similarity-based sampling is to pick core locations in such a way that the distribution of NDT measurements at these locations reproduces, as closely as possible, the distribution observed in the full NDT dataset [12]. To assess how close two distributions are, a variety of distance or similarity measures may be used, including the Cramér–von Mises (CvM) distance, the Earth Mover's Distance (EMD) and the Kolmogorov–Smirnov (KS) statistic. These metrics have been widely analysed and compared in different machine learning applications [14–16]. They all provide numerical indicators that summarise how two probability distributions differ from one another.

Depending on the problem, on practical constraints or on the nature of the data, one metric may be more appropriate than another. In the present study, the Kolmogorov–Smirnov statistic is selected as the distance measure used both to compare distributions and to quantify the degree of overlap between them.

Variance sampling:
The aim of the variance-based sampling method is to select a subset of NDT points that reflects as closely as possible the overall variability of the material. In other words, the selected samples should be statistically representative of the heterogeneity observed in the full dataset. To achieve this, a computational algorithm is used to systematically explore all possible combinations of NC core locations, where NC denotes the total number of available cores.

For each candidate combination, the corresponding NDT values are extracted and their variance is computed and compared with the variance of the complete dataset. The absolute difference between these two variances serves as the objective function to be minimised. By searching for the combination that minimises this difference, the method identifies an optimal set of core locations that best captures the spatial variability of the material across the structure.

Results and discussion
Assessing mean strength of the different sampling strategies:
To evaluate the accuracy and precision of each sampling strategy, the different approaches were applied to an experimental dataset. Both subfigures in Fig. 1 offer a comparative evaluation of different core sampling strategies, namely random, stratified, similarity-based, conditional, and variance-based, for estimating the mean compressive strength of concrete using non-destructive testing NDT data. Fig. 2. a) corresponds to the application of the bi-objective approach based solely on RH measurements, while Fig. 3. b) illustrates the multi-objective approach that combines RH and UPV measurements. In both figures, the y-axis represents the estimated mean strength (in MPa), and the x-axis shows the number of cores NC used for calibration. A reference line (dashed blue) indicates the experimental mean strength.

In Fig. 1(a), where only RH data are considered, the quality of the estimates is clearly very sensitive to the way cores are selected, especially for small values of NC (between 3 and 8). The random sampling scheme (red) tends to give the poorest results: its estimates of the mean strength remain offset from the reference value and exhibit noticeably larger fluctuations. By contrast, the variance-based sampling (blue) and the conditional sampling (yellow) lead to mean strength estimates that remain much closer to the reference, suggesting that the corresponding sets of cores are more representative of the material. Stratified and similarity-based sampling, although better than random selection in some cases, show more scattered behavior, particularly when NC is low, reducing their reliability.

In Fig. 4. b), where RH and UPV are combined in a multi-objective model, the impact of sampling strategies is less pronounced. All strategies, including random sampling, yield estimates that are generally closer to the reference value across the full range of NC. This suggests that the integration of multiple NDT techniques enhances the robustness of the estimation model, reducing sensitivity to the choice of core locations. However, even in this improved context, variance-based and conditional sampling continue to offer slight advantages, particularly in minimizing the variability of the estimated mean strength.

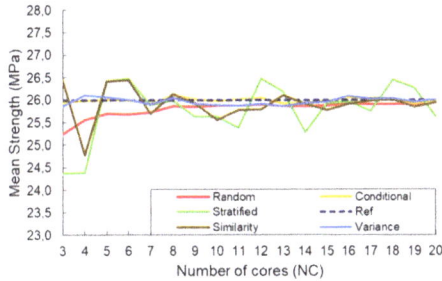

a) Rebound Hammer (bi-objective approach)

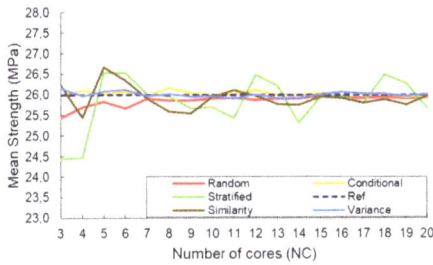

b) Combination of Rebound Hammer and Pulse Velocity (multi-objective approach)

Fig. 5 Mean strength estimation for the different sampling strategies: (a) based on Rebound Hammer measurements, and (b) based on the combined SonReb (UPV + RH) measurements.

Assessment of strength standard deviation of the different sampling strategies:
A meaningful evaluation of concrete performance does not only rely on the mean compressive strength; it must also take into account how much this strength varies from one point to another. Looking at the mean alone would mask a large part of the material heterogeneity. In this work, this aspect is handled through two identification frameworks: a bi-objective formulation when a single NDT method is used, and a multi-objective formulation when several NDT techniques are combined. Previous studies have shown that such formulations are well suited to estimating strength variability in a consistent way [17,18].

Figure 2 summarises how the estimated standard deviation of compressive strength evolves as the number of cores NC increases. The different sampling strategies are compared within the two modelling frameworks: the bi-objective model based solely on Rebound Hammer (RH) data (Fig. 2(a)), and the multi-objective model that jointly uses RH and UPV measurements (Fig. 2(b)). In both plots, the dashed black line ("Ref") corresponds to the reference variability computed from the full set of core tests.

In Fig. 2(a), the conditional sampling strategy (yellow curve) clearly tends to overestimate the variability, particularly when only a few cores are available. This behaviour is consistent with the way conditional sampling operates: it often selects locations associated with extreme NDT values, which leads to a sample that exaggerates the spread of strengths instead of representing the overall population. By contrast, the variance-based strategy (blue) and the similarity-based strategy (brown) display a much more regular trend and approach the reference standard deviation fairly quickly once NC exceeds about 10.

The random sampling strategy (red) shows a rather robust behaviour in both the bi-objective and multi-objective settings. Even for relatively small numbers of cores, the estimated variability remains close to the reference line and does not exhibit large deviations. The stratified sampling strategy (green) occupies an intermediate position. When only a few cores are considered, it tends to underestimate the actual variability, which can be attributed to an incomplete coverage of the material heterogeneity at that stage. As more cores are introduced, this initial bias is progressively corrected and the stratified estimates move towards the reference values.

In the multi-objective configuration illustrated in Fig. 2(b), where RH and UPV are combined, the contrast between the different sampling plans becomes less marked. The additional information provided by the second NDT technique stabilises the identification process and helps all sampling strategies converge towards the reference variability. This is a key observation for practical applications: when only a limited number of cores can be taken and the choice of their locations is constrained, the use of multiple NDT methods within a multi-objective framework can partly compensate for suboptimal sampling and still yield reliable estimates of strength variability.

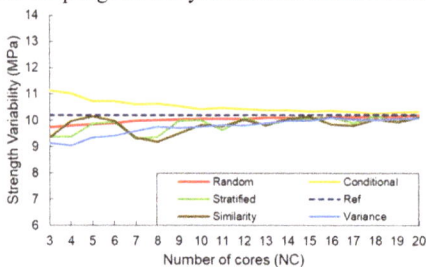

a) Rebound Hammer (bi-objective approach)

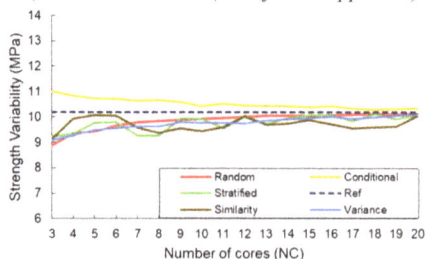

b) Combination of Rebound Hammer and Pulse Velocity (multi-objective approach)

Fig. 2 The strength variability estimated by a) Rebound Hammer, b) the combination of both techniques SonReb

Emerging Research in Materials for Environment, and Civil Infrastructure - GeoME 5.5   Materials Research Forum LLC
Materials Research Proceedings 58 (2026) 92-98   https://doi.org/10.21741/9781644903933-13

Conclusion

In the context of non-destructive control, the strategic selection of coring locations is critical for achieving a representative and precise evaluation of in-situ concrete strength. In order to choose appropriate location for the core extractions, the effectiveness of various proposed sampling methods has been assessed, namely variance-based, conditional, stratified, similarity-based and traditional random coring. We applied these strategies to a real experimental dataset and investigated the effect of single (RH) versus combined NDT techniques (SonReb) under both bi- and multi-objective calibration models. By systematically changing the number of cores, the sampling strategy and the NDT configuration, we were able to analyse how these three ingredients jointly affect the estimation of both the mean strength and its variability.

Two main observations can be drawn from the study. The first is that the choice of sampling strategy has a clear impact on the quality of the results. In particular, variance-based sampling leads to more accurate and more consistent strength estimates when only a small number of cores can be extracted. By guiding the selection towards more informative locations, these strategies reduce bias and make the predictions more trustworthy. The second observation is that using several NDT techniques within a multi-objective formulation further stabilises the assessment: the results become less sensitive to an imperfect or suboptimal choice of coring points.

From an engineering perspective, these results highlight that attention must be paid not only to the identification model, but also to the way cores are selected in the first place. The proposed framework can improve assessment quality without increasing coring costs, since the gains are obtained through better use of preliminary NDT measurements and more careful planning of the investigation. In practice, we therefore suggest incorporating variance-based sampling, preceded by an NDT survey of the structure, into routine assessment procedures, in order to improve both the accuracy and the cost-effectiveness of concrete strength evaluations in existing structures.

References

[1] K. Ali-Benyahia, Z.-M. Sbartaï, D. Breysse, S. Kenai, et M. Ghrici, « Analysis of the single and combined non-destructive test approaches for on-site concrete strength assessment: General statements based on a real case-study », *Case Stud. Constr. Mater.*, vol. 6, p. 109-119, juin 2017. https://doi.org/10.1016/j.cscm.2017.01.004

[2] M. Alwash, D. Breysse, et Z. M. Sbartaï, « Non-destructive strength evaluation of concrete: Analysis of some key factors using synthetic simulations », *Constr. Build. Mater.*, vol. 99, p. 235-245, nov. 2015. https://doi.org/10.1016/j.conbuildmat.2015.09.023

[3] B. Kouddane *et al.*, « Assessment of Concrete Strength Using the Combination of NDT—Review and Performance Analysis », *Appl. Sci.*, vol. 12, n° 23, p. 12190, janv. 2022. https://doi.org/10.3390/app122312190

[4] D. Breysse, D. Fokwa, et F. Drahy, « Spatial Variability in Concrete: Nature, Structure, and Consequences », *Appl. Mech. Rev.*, vol. 47, n° 1S, p. S184-S196, janv. 1994. https://doi.org/10.1115/1.3122812

[5] E. Saleh, A. Tarawneh, H. Dwairi, et M. AlHamaydeh, « Guide to non-destructive concrete strength assessment: Homogeneity tests and sampling plans », *J. Build. Eng.*, vol. 49, p. 104047, mai 2022. https://doi.org/10.1016/j.jobe.2022.104047

[6] K. Ali-Benyahia, Z.-M. Sbartaï, D. Breysse, M. Ghrici, et S. Kenai, « Improvement of nondestructive assessment of on-site concrete strength: Influence of the selection process of cores location on the assessment quality for single and combined NDT techniques », *Constr. Build. Mater.*, vol. 195, p. 613-622, janv. 2019. https://doi.org/10.1016/j.conbuildmat.2018.10.032

[7] D. Breysse *et al.*, « Non destructive assessment of in situ concrete strength: comparison of approaches through an international benchmark », *Mater. Struct.*, vol. 50, n° 2, p. 133, févr. 2017. https://doi.org/10.1617/s11527-017-1009-7

[8] D. Breysse, X. Romão, M. Alwash, Z. M. Sbartaï, et V. A. M. Luprano, « Risk evaluation on concrete strength assessment with NDT technique and conditional coring approach », *J. Build. Eng.*, vol. 32, p. 101541, nov. 2020. https://doi.org/10.1016/j.jobe.2020.101541

[9] K. Ali Benyahia, Z. M. Sbartaï, B. Denys, S. Kenai, et M. Ghrici, « Analysis of the single and combined non-destructive test approaches for on-site concrete strength assessment: General statements based on a real case-study », *Case Stud. Constr. Mater.*, vol. 6, janv. 2017. https://doi.org/10.1016/j.cscm.2017.01.004

[10] K. Ali-Benyahia, Z.-M. Sbartaï, D. Breysse, M. Ghrici, et S. Kenai, « Improvement of nondestructive assessment of on-site concrete strength: Influence of the selection process of cores location on the assessment quality for single and combined NDT techniques », *Constr. Build. Mater.*, vol. 195, p. 613-622, janv. 2019. https://doi.org/10.1016/j.conbuildmat.2018.10.032

[11] W. G. Cochran, *Sampling Techniques, 3rd Edition*. New York, 1977. Consulté le: 4 août 2023. [En ligne]. Disponible sur: https://www.wiley.com/en-us/Sampling+Techniques%2C+3rd+Edition-p-9780471162407

[12] E. Saleh, A. Tarawneh, H. Dwairi, et M. AlHamaydeh, « Guide to non-destructive concrete strength assessment: Homogeneity tests and sampling plans », *J. Build. Eng.*, vol. 49, p. 104047, mai 2022. https://doi.org/10.1016/j.jobe.2022.104047

[13] J.-F. Wang, A. Stein, B.-B. Gao, et Y. Ge, « A review of spatial sampling », *Spat. Stat.*, vol. 2, p. 1-14, déc. 2012. https://doi.org/10.1016/j.spasta.2012.08.001

[14] S. Boriah, V. Chandola, et V. Kumar, « Similarity Measures for Categorical Data: A Comparative Evaluation », in *Proceedings of the 2008 SIAM International Conference on Data Mining (SDM)*, in Proceedings. , Society for Industrial and Applied Mathematics, 2008, p. 243-254. https://doi.org/10.1137/1.9781611972788.22

[15] D. J. Weller-Fahy, B. J. Borghetti, et A. A. Sodemann, « A Survey of Distance and Similarity Measures Used Within Network Intrusion Anomaly Detection », *IEEE Commun. Surv. Tutor.*, vol. 17, n° 1, p. 70-91, 2015. https://doi.org/10.1109/COMST.2014.2336610

[16] J. Irani, N. Pise, et M. Phatak, « Clustering Techniques and the Similarity Measures used in Clustering: A Survey », *Int. J. Comput. Appl.*, vol. 134, n° 7, p. 9-14, janv. 2016. https://doi.org/10.5120/ijca2016907841

[17] B. Kouddane, Z. M. Sbartaï, S. M. Elachachi, et N. Lamdouar, « New multi-objective optimization to evaluate the compressive strength and variability of concrete by combining non-destructive techniques », *J. Build. Eng.*, vol. 77, p. 107526, oct. 2023. https://doi.org/10.1016/j.jobe.2023.107526

[18] Z.-M. Sbartaï, M. Alwash, D. Breysse, X. Romão, et V. A. M. Luprano, « Combining the bi-objective approach and conditional coring for a reliable estimation of on-site concrete strength variability », *Mater. Struct.*, vol. 54, n° 6, p. 230, nov. 2021. https://doi.org/10.1617/s11527-021-01820-9

Emerging Research in Materials for Environment, and Civil Infrastructure - GeoME 5.5   Materials Research Forum LLC
Materials Research Proceedings 58 (2026) 99-107   https://doi.org/10.21741/9781644903933-14

Influence of natural additives on the heat transfer properties of clay–plaster building materials

Soufian Omari[1,a] *, Najma Laaroussi[1,b] and Aziz Ettahir[1,c]

[1]Mohammed V University in Rabat, Materials, Energy, and Acoustics Team (MEAT), Higher School of Technology of Salé, Morocco

[a]soufianmouhamed15@gmail.com

Keywords: Energy Efficiency, Clay, Gypsum Plaster, Thermal Performance, Sustainable Materials

Abstract. The energy efficiency of the batteries passes through the use of environmental protection materials. In Morocco, a paid sub-stitut is located on a gypsum base and argile. The gypsum plate with good quality thermal insulation (0.2-0.35 W/m K) tandis that the argile content stabilizes the indoor grate temperature with forte thermal insulation (0.4-1.2 W/m K) and its environmental impact. The association of these materials with natural additives allows for the reduction of CO_2 emissions, the comfort of the interior and the lower levels of chafing and climate from 30 to 50%. Integration of traditional materials and modern construction techniques favors the development of durable constructions, adaptations to local climates and a feasible carbon construction.

1. Introduction

The high energy consumption of the building sector is a major challenge, placing considerable pressure on financial and energy resources, particularly in developing countries such as Morocco.

The heating, cooling, and lighting systems in buildings account for a large amount of energy consumption. Using technologies that preserve thermal comfort while enabling energy reduction is becoming essential. Plâtre and argile are acknowledged as efficient and eco-friendly materials due to their thermal characteristics. For ages, argile has been utilized to decrease or even eliminate the need for heating and mechanical climate control because of its thermal inertie, which stores heat during the day and recovers it at night. It is long-lasting, reusable, and simple to produce. Its thermal diffusivity is between 0.1 and 0.3 mm²/s, and its thermal conductivity is between 0.25 and 1.5 W/m·K. Plâtre completes the argile in walls and panels because to its low thermal conductivity (0,16–0.4 W/m·K) and natural fire resistance. Argile-plâtre composites improve insulation, durability, and humidity management. They also help to keep indoor temperatures and humidity levels steady, save energy use, and make indoor environments healthier.

Morocco possesses vast amounts of gypsum and clay, which have tremendous economic potential. Thousands of people are directly employed in the gypsum industry, and clay continues to be an essential part of both traditional and modern construction. Making use of these domestic resources strengthens the economy and lessens reliance on imports.As the building industry moves toward sustainability, clay-gypsum composite systems offer a workable means of reducing energy consumption, carbon emissions, and encouraging environmentally friendly building practices. In order to increase the applications of clay-based materials and promote sustainable building practices, this study builds on earlier research on the subject. The energy performance of clay composites was compared to that of common building materials in a range of climates by comparative testing.The specific studies include, for example: Energy performance comparisons between clay-based materials and traditional materials. Alioui et al.,[1]. Clay–straw composites. Azhary et al., [2]; Hemp fibers in plaster Charai et al.,[3]. Cork-plaster composites. Cherki et al.,[4]. Cow dung in clay bricks. Dadi Mohamed et al.,[5]. clay brick mixtures. Laaroussi et al.,[6].

Emerging Research in Materials for Environment, and Civil Infrastructure - GeoME 5.5   Materials Research Forum LLC
Materials Research Proceedings 58 (2026) 99-107   https://doi.org/10.21741/9781644903933-14

Peanut shell plater composites. Lamrani et al.,[7]. Clay and olive pruning waste Liuzzi et al.,[8]. hemp–lime plasters. Mazhoud et al.,[9]. Gypsum–clay–sand composites Omari et al.,[10]. Collectively, these studies establishes the merits of utilizing natural, low-cost additives that optimize thermal performance characteristics that leads to improved energy performance and supports sustainable strategies in building.

2. Materials and Methods

2.1 Sample Preparation

Samples were made using a traditional procedure that is still used in the south of Morocco. To ensure total moisture saturation, raw soil was blended with 19% water by dry weight and allowed to rest for three days. Straw fibers were then added in varying amounts of 0%, 2%, and 4% before forming the adobe bricks. The shaped samples were then allowed to dry in natural ambient conditions for 28 days [1].

Clay with a 22% moisture content was used to make Slaoui hollow bricks, which were first dried for three hours at 110°C and then fired for twenty-four hours at 780°C. The drying process is more environmentally friendly and energy efficient since it makes use of the kiln's waste heat. This chauffage rends the briquet, which is more solid, more durable, and comes out against the chaleur. The results are based on the existing minerals in the argile and the façon is not the same [6].

Because they effectively regulate heat, these bricks help buildings use less energy. They are a lightweight, durable material that is simple to handle and install, making them appropriate for energy-efficient, sustainable buildings in any climate (Table 1).

Table 1. Dimensions of the sample

Dimension (mm^3)	100x100x26
Mass (g)	462,14
Density (kg/m^3)	1777

Several research have compared the structural and thermal qualities of clay-straw fiber composites and regular clay bricks. In one study, straw fibers (1.25-2.5 mm in diameter) were mixed with clay at different volumes (E1-E5) while maintaining a constant water-to-clay ratio of 0.23. Standard dimensions: 100 x 100 x 22 mm.^{3} samples were dried at 60°C to eliminate moisture and prevent deformation [2]. The study aimed to evaluate the impact of straw fiber content on mechanical strength, thermal insulation, and durability, which are important properties of sustainable materials.

Another study assessed four clay-based plaster formulations made of clay, sand, gravel, and leftovers from olive branches that contained olive fibers (about 3 cm). The clay fraction had 71% clay, 28% silt, 22% carbonates, and 1% sand. Water, hydrated lime, quartzite gravel (2-4 mm), quartzite sand (<2 mm), and quarry fines made up the soil matrix. After 28 days of curing, olive fibers improved their mechanical and thermal capabilities, suggesting that they could be used in ecologically friendly construction [8], (Table 2).

Table 2. *The formulation of the matériel*

Code mix	Gravel %.	Clay %.	Sand%.	Olive fiber%.
4	2	38	56	4
6	2	38	54	6
8	2	37	53	8
12	2	37	51	12

In Abéché, around 900 kilometers east of N'Djamena, this study aims to advance sustainable building practices by using natural materials into plaster composites and clay bricks. Clay bricks were molded in a $100 \times 100 \times 50$ mm³ mold after cow dung was added at a rate of 0-5%. After adding expanded perlite and cork grains (6.3–8 mm) in different proportions (0–100%), the mixture was poured into a $100 \times 100 \times 20$ mm³ mold. After being dried in a kiln to a constant mass, the samples' mechanical strength, thermal insulation, and performance were assessed [4].

A 50/50 mixture of gypsum plaster (GP) and pottery clay (PC) was mixed with sand at a 5–20% ratio. Before being used, the cleaned sand was sieved to a particle size of 6.3 to 8 mm and left to air dry for 72 hours.

Samples measuring 100 x 100 x 20 mm3 were created with a water-to-gypsum ratio of 0.7, air-dried for 24 hours, then baked for 48 hours at 50°C. Each sample was sealed in plastic bags to prevent moisture absorption, and the inclusion of sand improved and hastened the drying process, particularly in gypsum-based mixtures [10].

We evaluated plaster compositions containing 2.5-10 mm peanut particles. In one case, we used 20% peanuts and 0-20% plaster, while in another, we used a range of peanut sizes, 100% cork, and 20-80% plaster. Samples were cut into $100 \times 100 \times 20$ mm³ cubes and dried in both vacuum and air. Hemp fibers (10-20 mm) from northern Morocco were mixed with gypsum at a water-to-plaster ratio of 0.5, gently crushed, and allowed to cure for 48 hours in 150×150 x 20 mm molds. Levels of fiber ranged from 0% to 6% [9].

In order to better understand clay and plaster composites reinforced with cow dung, perlite, cork, peanut shell particles, and hemp fibers for long-term high-performance structures, the study examined their mechanical, thermal, and hydraulic properties.

2.2 Samples preparation

2.2.1 Thermal Conductivity

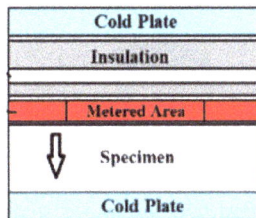

Fig. 1. Guarded Plate Technique

One popular approach for determining a material's thermal conductivity in steady-state conditions is the shielded hot plate method. This technique creates a temperature gradient that allows for an accurate measurement of the heat flux through the sample by sandwiching a sample between two parallel plates, one of which is heated and the other at a reference temperature. Following the achievement of thermal equilibrium, the temperature differential is noted, and the

Emerging Research in Materials for Environment, and Civil Infrastructure - GeoME 5.5   Materials Research Forum LLC
Materials Research Proceedings 58 (2026) 99-107   https://doi.org/10.21741/9781644903933-14

measured heat flux is used to compute the thermal conductivity. Owing to its high accuracy, this method is commonly used in laboratories for materials with moderate conductivity, as shown in (Fig. 1). In practice, heating elements are positioned between the sample and insulating layer, while sensors on the unheated surfaces of the sample and insulating foam record temperatures T_1 and T_2. The heat flux densities through the sample and insulation (f_1 and f_2) are then determined for precise thermal conductivity evaluation, (Eq. 1, Eq. 2).

$$\Phi_1 = \frac{T_0 - T_1}{\lambda_1 \, e_1} \tag{1}$$

$$\Phi_2 = \frac{T_2 - T_1}{\lambda_2 \, e_2} \tag{2}$$

Where :
T_0 indicates the thermal state of the energized face, T_1 the temperature at the unexposed surface of the sample, and T_2 the temperature of the non-energized side of the foam. The conductive properties of the sample and foam are λ_1 and λ_2, while their respective layer dimensions are e_1 and e_2. The overall energy flux (Φ_0) is influenced by the applied voltage (U), the heater's resistance (R), and the sectional area (S). (Eq. 3):

$$\emptyset = \frac{U^2}{R.S} \tag{3}$$

Combining the heat flux equations yields the following equation for determining the sample's thermal conductivity (λ_1) (Eq. 4) :

$$\lambda_1 = \frac{e_1}{((T_0 - T_1))} = \left[\frac{U^2}{R.S} - \frac{\lambda_2}{e_2} (T_0 - T_2) \right] \tag{4}$$

This equation is used to calculate thermal conductivity when the system has reached steady-state conditions. This approach gives reliable readings since it takes into account contributions from both the material sample and the insulating foam. It has been effectively used to a wide variety of materials and is still one of the most often used methods for evaluating and comparing thermal performance.

2.2.2 Thermal diffusivity

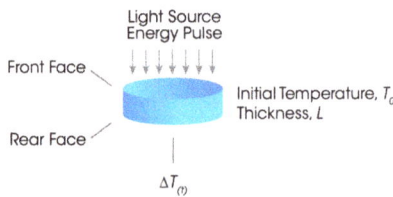

Fig. 2. Thermal Diffusivity Measurement via the Flash Technique

The Transient Heat Pulse Method, shown in (Fig. 2), is widely employed to measure the rate of heat propagation in solid materials. In this technique, a brief heat pulse is applied to one surface of a thin sample, and the resulting thermal response, T(t), is recorded on the opposite face. The thermal diffusivity (α) of a material can be evaluated by studying its transient temperature profile, which indicates the rate of heat movement. This attribute is directly related to the material's thermal conductivity (λ), density (ρ), and specific heat capacity (Cp), as shown in the following connection, (Eq. 5):

$$\alpha = \frac{0.1388 \cdot L^2}{t_{1/2}} \tag{5}$$

Where :

α is the thermal diffusivity (in m^2/s), L is the sample thickness (in meters), and $t_{1/2}$ is the time needed for the rear face of the sample to attain 50% of its maximum temperature rise after the heat pulse (in seconds). The constant 0.1388 is a dimensionless value calculated from an analytical solution to the heat equation under ideal conditions.

3. Results and discussions

3.1 Results

Adding wheat straw to the clay samples during the first phase dramatically decreased their density. The control bricks (B0) had a density of 1936 kg/m^3, whereas those with 2% and 4% straw had densities of 1703 kg/m^3 and 1614 kg/m^3, respectively, suggesting a reduction of 12% and 16.3%. This drop in density had a direct effect on the thermal behavior of the composites. Bricks containing 2% and 4% straw reduced thermal conductivity by 29.65% and 49.12%, respectively [1].

The clay samples' density was considerably decreased in the first phase when wheat straw was added. Bricks containing 2% and 4% straw had densities of 1703 kg/m^3 and 1614 kg/m^3, respectively, indicating a reduction of 12% and 16.3% from the control bricks (B0), which had a density of 1936 kg/m^3. The thermal behavior of the composites was directly impacted by this density decrease. The thermal conductivity of the bricks made with 2% and 4% straw decreased by 29.65% and 49.12%, respectively [2], (Table 3).

Table 3. Density and Thermal Properties of Clay–Straw Composite Samples

Samples	$\rho(Kg/m^3)$	$\lambda(W/m\ K)$	$a(m^2/s)\ x10^{-7}$	$Cp(KJ/kg.K)$
B0	1936	0,575	3,837	0,774
B2	1703	0,401	3,312	0,752
B4	1614	0,294	2,451	0,743

Using a three-dimensional heat conduction model with boundary conditions at y = 0 and y = 2C + e, it was found that lowering thermal conductivity increases thermal resistance and limits heat exchange with the surroundings. The brick slab has a volumetric heat storage capacity of $\rho C = 4.8 \times 10^3$ $J/m^3 \cdot K$ and a mean thermal conductivity of $\lambda_m = 0.346$ W/m·K, according to experimental results. The efficiency with which a material absorbs, stores, and transfers thermal energy is determined by these characteristics [6].

Moroccan slate brick is a superb option for engineering and building because of its exceptional strength and thermal properties. Three distinct thermal settings were used to test the clay-straw composite's energy performance in order to better understand how it reacts to temperature fluctuations [2], (Table 4, Table 5).

Table 4. Thermal Conductivity Test Results

Sample	Straw %.	$\lambda 1$	$\lambda 2$	$\lambda 3$	$\lambda 4$
E5	5	0,266	0,265	0,260	0,263
E4	4	0,305	0,306	0,309	0,306
E3	3	0,356	0,360	0,358	0,358
E2	2	0,407	0,400	0,409	0,405
E1	0	0,508	0,505	0,510	0,504

Table 5. *Test results: Thermal diffusivity (m² /s) x10⁻⁷*

Sample	Straw %.	a_1	a_2	a_3	a_4
E5	5	2,875	2,818	2,881	2,858
E4	4	3,119	3,125	3,092	3,112
E3	3	3,393	3,393	3,405	3,306
E2	2	3,521	3,518	3,486	3,508
E1	0	3,727	3,752	3,705	3,728

Straw improves insulation and limits heat transfer by lowering thermal conductivity and diffusivity when added to the clay-straw composite. The material is particularly well-suited for construction applications as a result. These composites are especially effective at lowering summer heat and winter cold in arid and semi-arid regions, improving indoor comfort.

Four combinations containing different amounts of clay, sand, gravel, olive fibers, and water were made in order to gain a better understanding of this influence. Each sample was tested for thermal reactivity in a range of conditions to ascertain the effects of each component on thermal conductivity and diffusivity. The results of the experiment offer vital information for maximizing natural building mixtures in locations with significant temperature fluctuations [8], (Table 6).

Table 6. *Thermal Variations Test Results*

Mixture	Density(kg/m³)	λ(W/m.k)	P(J/kg.K)	a(m² /s) x10⁻⁶
4	1669	0,593	849,6	0,387
6	1599	0,532	869,7	0,365
8	1497	0,458	908,6	0,320
12	1409	0,428	958,9	0,295

Adding olive fibers to a mixture lowers its heat conductivity, which affects other characteristics like diffusivity. These alterations reveal the remarkable thermal qualities of the clay-olive fiber blend, which make it a desirable alternative material for green construction.

Cow dung reduced specific heat capacity (Cp_1 and Cp_2), indicating improved insulation and a change in the material's thermal properties. Composites with high strength and low thermal conductivity improve energy efficiency and lessen the effects of temperature changes.

One locally accessible natural supplement that improves material performance and promotes environmental sustainability is cow dung. These results show the advantages of employing inexpensive, natural materials for environmentally friendly and energy-efficient building, but they also show that more research is necessary to fully realize their potential [5], (Table 7).

Table 7. *Test results of the mixture materials CP₁ and CP₂*

Sample	Cow dung	Cp_1(J/kg/k)	Cp_2(J/kg/k)
E0	0%	871	885
E1	1%	736	745
E2	2%	708	723
E3	3%	707	708
E4	4%	689	697
E5	5%	677	693

Table 8. Thermal Conductivity of Samples

No.	Samples	λ (W/mK)
E1	CP (100%)	0,448
E1	GP (100%)	0,454
E2	CP (50%) +GP (50%)	0,426
E3	CP (50%) +GP (50%) +SA (5%)	0,440
E4	CP (50%) +GP (50%) +SA (10%)	0,472
E5	CP (50%) +GP (50%) +SA (20%)	0,484

Thermal qualities are stressed while evaluating construction materials for energy-efficient structures. Because of their comparable thermal conductivities, gypsum plaster (GP) and pottery clay (CP) can be used interchangeably in building. When mixed equally, the composite exhibits a lowered conductivity of 0.426 W/m·K, indicating positive microstructural interactions that improve insulation. Adding 5-20% sand (SA) reduces thermal resistance marginally and increases conductivity to 0.484 W/m·K. However, the GP-CP mixture outperforms regular clay bricks (about 0.65 W/m·K). This makes it an economical and weather-appropriate material for the area of Rabat, Salé, and Kénitra [10] (Table 8).

Additionally, the impact of peanut shell particles (6.3–8 mm, 0–20%) in Ps–Pl plaster composites was examined. Granular cork was found to have a high insulating potential because it reduced conductivity while increasing thermal diffusivity from 2.12 to 3.75×10^{-7} m²/s and conductivity from 0.141 to 0.301 W/m·K [7]. The correctness of these findings was confirmed by complementary approaches: the evaluations of convective heat transfer and heat flow showed variances of 5% and 9%, respectively, while the Parker and Degiovanni methods for thermal diffusivity had a maximum error margin of 3%. These findings demonstrate that natural additives can increase thermal efficiency and that the employed techniques are workable (Table 9).

Researchers have examined chanvre-based isolant materials with densities ranging from 720 to 880 kg/m³. Denser materials provide better insulation, reduce heat transfer, and help buildings use less energy. Additionally, the chanvre helps to regulate humidity, enhance ventilation, and prevent moisture buildup, all of which contribute to healthier indoor air. These features are beneficial for the passengers' comfort and temperature. Because of its strength, usefulness, and environmental friendliness, hemp is a great substitute for conventional building techniques. Achieving energy performance and environmental sustainability goals is made achievable by optimizing density and utilizing natural resources [3].

Table 9. Two types of thermal measurement results.

Hot wire		P (Kg/m³)	Hot disk λ (W/m/K)	a (mm² /s)
λ(W/m/K)			0,1795	0,2466
CHLP Average value 733 0.1963			0,0099	0,0046
Standard deviation 7 0.0044			2,6 4	
Var Coef 1 2,2			0,2277	0,2142
THLP Average value 887 0.2364			0,0066	0,0057
Standard deviation 14 0.0082			2,9 2,7	
Var Coef 1,6 3,5				

Emerging Research in Materials for Environment, and Civil Infrastructure - GeoME 5.5   Materials Research Forum LLC
Materials Research Proceedings 58 (2026) 99-107   https://doi.org/10.21741/9781644903933-14

Table 10. *Thermal measurement results.*

Pourcentage de fibre %	λ(W/m.K)	a(mm² /S)	Effusivité thermique (W.$s_{1/2}$/m². K)	Vol. Spécifique (MJ/m². K)
0	0,531±0,001	0,400±0,011	839±09	1,327±0,032
2	0,485±0,001	0,390±0,008	776±07	1,243±0,025
4	0,441±0,003	0,381±0,019	714±13	1,159±0,050
6	0,364±0,001	0,368±0,003	601±03	0,990±0,009

Plaster's insulating qualities are greatly enhanced by hemp, which lowers thermal conductivity from 0.531 to 0.364 W/m·K and thermal diffusivity from 0.399 to 0.365 mm²/s while raising thermal resistance by 31.5%. Additionally, there is a about 24.8% decrease in the volumetric heat capacity and a 27.3% decrease in the heat exchange capacity [10]. These modifications demonstrate how hemp might improve building materials' energy efficiency, (Table 10).

4. Discussion

These materials help to maintain consistent indoor temperatures and humidity levels, which enhances comfort and lowers energy use. The plâtre absorbs and releases moisture, and with proper ventilation, it helps control the temperature, reduce the need for heating or air conditioning, and prevent moisture buildup. On the other hand, the argile absorbs heat during the day and releases it at night, making it an effective natural insulator. L'ajout de fibres végétales comme le lin, le chanvre ou la paille améliore encore l'isolation, la résistance et la maniabilité du matériau. Ces fibres, légères, biodégradables et renouvelables, réduisent les besoins en énergie et les émissions de gaz à effet de serre. En privilégiant des matériaux locaux, on diminue aussi l'impact environnemental tout en soutenant l'économie. Les mélanges d'argile et de plâtre renforcés par des fibres naturelles offrent ainsi une solution écologique et performante pour les bâtiments modernes.

Conclusions

An increasing number of people want homes that use less energy and are less expensive to maintain. Reducing expenses and energy consumption is necessary for good thermal comfort. Increasing the quality of building materials is a good answer. Plâtre and argile are excellent isolants that help maintain a constant temperature and humidity, which reduces the need for heating or air conditioning. Adding natural fibers like paille or chanvre enhances isolation even more. These environmentally friendly materials enable the construction of more beautiful, cost-effective, and environmentally friendly buildings.

References

[1] A. Alioui, Y. Azalam, M. Benfars, E.M. Bendada, M. Mabrouki, Comparative analysis of energy performance between clay-based and conventional building materials: A case study in Moroccan semi-arid climate, E3S Web Conf. 582 (2024) 01005. https://doi.org/10.1051/e3sconf/202458201005

[2] K.E. Azhary, Y. Chihab, M. Mansour, N. Laaroussi, M. Garoum, Energy Efficiency and Thermal Properties of the Composite Material Clay-straw, Energy Procedia 141 (2017) 160–164. https://doi.org/10.1016/j.egypro.2017.11.030

[3] M. Charai, H. Sghiouri, A. Mezrhab, M. Karkri, Thermal insulation potential of non-industrial hemp (*Moroccan cannabis sativa L.*) fibers for green plaster-based building materials, Journal of Cleaner Production 292 (2021) 126064. https://doi.org/10.1016/j.jclepro.2021.126064

[4] A. Cherki, A. Khabbazi, B. Remy, D. Baillis, Granular Cork Content Dependence of Thermal Diffusivity, Thermal Conductivity and Heat Capacity of the Composite Material/Granular Cork Bound with Plaster, Energy Procedia 42 (2013) 83–92. https://doi.org/10.1016/j.egypro.2013.11.008

[5] A.D. Mahamat, O.I. Hamid, M. Soultan, M.Y. Khayal, Y. Elhamdouni, M. Garoum, S. Gaye, Effect of Cow's Dung on Thermophysical Characteristics of Building Materials Based on Clay, Research Journal of Applied Sciences, Engineering and Technology 10 (2015) 464–470. https://www.airitilibrary.com/Article/Detail/20407467-201406-201507080016-201507080016-464-470 (accessed October 3, 2025).

[6] N. Laaroussi, A. Cherki, M. Garoum, A. Khabbazi, A. Feiz, Thermal Properties of a Sample Prepared Using Mixtures of Clay Bricks, Energy Procedia 42 (2013) 337–346. https://doi.org/10.1016/j.egypro.2013.11.034

[7] M. Lamrani, N. Laaroussi, A. Khabbazi, M. Khalfaoui, M. Garoum, A. Feiz, Experimental study of thermal properties of a new ecological building material based on peanut shells and plaster, Case Studies in Construction Materials 7 (2017) 294–304. https://doi.org/10.1016/j.cscm.2017.09.006

[8] S. Liuzzi, C. Rubino, P. Stefanizzi, Use of clay and olive pruning waste for building materials with high hygrothermal performances, Energy Procedia 126 (2017) 234–241. https://doi.org/10.1016/j.egypro.2017.08.145

[9] B. Mazhoud, F. Collet, S. Pretot, J. Chamoin, Hygric and thermal properties of hemp-lime plasters, Building and Environment 96 (2016) 206–216. https://doi.org/10.1016/j.buildenv.2015.11.013

[10] S. Omari, N. Laaroussi, A. Ettahir, The influence of adding sand to a composite material based on gypsum plaster and pottery clay on thermal conductivity, AIP Conference Proceedings 3241 (2024) 020004. https://doi.org/10.1063/5.0247011

Emerging Research in Materials for Environment, and Civil Infrastructure - GeoME 5.5   Materials Research Forum LLC
Materials Research Proceedings 58 (2026) 108-115   https://doi.org/10.21741/9781644903933-15

New challenges in the field of heritage conservation and restoration

Daniela Pittaluga[1,*]

[1]Dipartimento Architettura e Design, Università degli Studi di Genova, Italy

daniela.pittaluga@unige.it

Keywords: Digital Technologies, Cultural Heritage, Restoration, Mensiocronology, BIM

Abstract. The problems of continuously growing world population, progressive consumption of natural resources, environmental challenges related to climate change, challenge the entire Mediterranean world. In Italy attempts have been made for some time now to provide answers to this continuous and progressive consumption of resources by seeking sustainable solutions from various points of view: zero soil consumption is one of the initiatives put in place. This in- volves directing more attention towards the historic built environment. Howev- er, how can interventions be carried out in an environmentally sustainable man- ner while respecting the specific characteristics of historic buildings? How to preserve the uniqueness of traditional building systems and that we through res- toration work should pass on to the future? How can structures in old buildings be made safe from a static point of view without disrupting their structural de- sign? For years, the University of Genoa have been collaborating on these is- sues with the CNR and University of Florence. Recent research has shown how some digital tools and technologies can help on the monitoring of the state of preservation. This enables timely action, lowering intervention costs and im- proving heritage conservation. Other research developed at national level on BIM systems, have shown that in the field of historical buildings and restora- tion, specific expedients are necessary with respect to ex novo interventions. All this gives rise to the need for more collaboration in the future between experts with knowledge of the historic built heritage, experts in diagnostics and materi- als scientists and experts in digital technologies and artificial intelligence to be able to intervene on historic buildings in a truly sustainable manner.

1. Introduction

The problems highlighted (continuously growing world population, continuous and progressive consumption of natural resources, environmental challenges related to climate change) challenge the entire Mediterranean world. In Italy in particular, at- tempts have been made for some time now to provide answers to this continuous and progressive consumption of resources by seeking sustainable solutions from various points of view: zero soil consumption is one of the initiatives put in place. This involves directing more attention towards the historic built environment. But how to intervene in an environmentally sustainable way respecting the peculiarities of histor- ic buildings? How to preserve the uniqueness of traditional building systems that have come down to us and that we through restoration work should pass on to the future? In what ways can the static stability of historic buildings be ensured while preserving their structural integrity and design? For years, the University of Genoa, in particular the DAD Department of Architecture and Design and the Department of Engineering (DICCA - Department of Civil, Chemical and Environmental Engineering) have been collaborating with research on these issues also with the CNR and the University of Florence. Recent research has shown how some digital tools and technologies can help the monitoring of the state of preservation. This makes it possible to intervene in a timely manner, reducing the costs of interventions and, on the other hand, allowing greater conservation of historical heritage. Other research on BIM systems, with research projects also developed at the national level, have shown that in the field of historical buildings and restoration, specific strategies are necessary compared to new constructions.

Emerging Research in Materials for Environment, and Civil Infrastructure - GeoME 5.5   Materials Research Forum LLC
Materials Research Proceedings 58 (2026) 108-115   https://doi.org/10.21741/9781644903933-15

2. Materials, Method, and Discussion

Digital archiving systems, BIM increasingly constitute support in the construction field. But is this also true in the field of Restoration and Conservation? Or are other tools necessary when working on historic buildings? Are special precautions necessary? Are the simplifications imposed by various modelling programs impractical for those who must ensure the conservation of Monumental and Non-Monumental Heritage? These are the main questions that were the starting point for a PRIN Project, a Project of National Significance PRIN 2010-2011, *"Built Heritage Information Mod- elling/Management - BHIMM"*, scientific coordinator S. Della Torre, Politecnico di Milano (code 20104TA4Y8); this project started in 2013, ended in 2016 and investigates the possibility of using BIM (Building Information Modelling) technology to the intervention process on existing and, moreover, monumental artefacts. When the PRIN project was launched, the literature on the application of digital information models to historic buildings was still very limited [e.g.1,2]. Just five years later, however, significant progress had been made in this field of research, as demonstrated by numerous studies [3,4,5]. These studies are still viewed with interest today, and new research is ongoing. BIM applied to cultural heritage, or Historic BIM, seems to have become an increasingly attractive topic for academic research, but it has also brought up many problems and doubts. BIM modelling applied to cultural heritage must address specific issues arising from conflicting requirements. On the one hand, model- ling is, by its very nature, a schematic and simplified representation of the object under study; on the other hand, restoration requires the creation of complex, accurate and detailed models. Finding a balance between these two approaches is not easy, as each historic building has unique characteristics that require a customised working method. It is therefore essential to define clear criteria for determining which elements to represent and with what minimum level of detail. In the field of restoration, it is essential to enhance the uniqueness of each piece of architecture, describing its most distinctive parts with precision and consistency. In a BIM context, modelling cultural heritage can follow two contrasting paths: detailed accuracy or schematic abstraction. Two operational possibilities that meet different objectives with different tools. From the use of BIM authoring programs, such as Revit or ArchiCAD, it emerges how difficult it is to reconcile the uniqueness of the elements that characterize the architectural work and the strong standardization of the parametric objects inserted in the libraries of this type of software. To overcome this problem, one could think of reducing the architectural elements to the typified objects in the library, inevitably losing precision in the restitution of the model. However, given that in the field of cultural heritage interest in detailed knowledge of the artefact has led to the development of surveying techniques and restitution of the data with ever-increasing precision, the solution of reducing the specific architectural elements to standardized elements does not seem satisfactory [6].

2.1 BIM and Archeology of Architecture e la sperimentazione all'Albergo dei Poveri (Genova, ITALY)

Other aspects investigated in the Prin included assessing the possible application in BIM of the method of investigating the archaeology of architecture; method already widely used in the field of Architectural Restoration [7]. What is the Archaeology of Architecture? It is a method of knowledge of historical constructions; it derives from archaeological methods. How can it be useful in the Restoration of buildings? Provides a good knowledge of the historical structure and its transformations over time and in this sense may have a relationship with Bim. Since the 1970s, numerous non-destructive tools have been created or perfected for the archaeological study of the still existing and visible architecture of our historic centres. In Genoa (Italy) innovative experiences have been carried-on by ISCUM (Istituto di Storia della Cultura Materiale – Genova – Italy) and by the "Laboratorio di Archeologia dell' Architettura" (University of Genoa) [8, 7]. Thanks to this research, the tools described below have been made accessible in Genoa and throughout the Liguria region. At the same time, other teams are adapting them for use in different areas in Italy and in different parts of Europe (and beyond), which means that it is now possible to apply a wide range of analytical and non-invasive techniques for the archaeological study of historical architecture across a large area.

Building stratigraphy is a technique derived from archaeological excavation, used to reconstruct the sequence of construction and destruction that has led to the building's current state. It is based on the concept of stratigraphic units (SU), i.e. parts of the building constructed in a single phase of work. Over time, positive (additions) and negative (demolitions) SUs have overlapped, some preserved, others ruined or disappeared. Through observation and analysis of the points of contact between the SUs, the stratigraphic sequence is defined, allowing relative (not absolute) chronological relationships between the construction phases to be established. This method provides a stable basis on which to integrate absolute dating obtained using other techniques, avoiding misinterpretations that would result from isolated data. [8]. Chronotypology is a dating method that compares architectural elements with a database of dated examples, linking forms and construction techniques to specific historical periods. It is based on statistical analysis and only works in homogeneous geographical and cultural contexts. Dendrochronology:(from the Greek "déndron", meaning "tree") is a method of dating wooden artefacts based on the analysis of variations in tree growth rings. It is a non-destructive method when it is possible to observe a sufficient number of rings on the section of wood to be analysed, or microdestructi when it is necessary a sample. This technique is widely used in archaeology, and also in architectural archaeology, to determine the age of wooden materials used in construction. For different wood species and in various erritorial contexts, dendrochronological curves have been developed, based on samples of certain age. These curves relate the growth characteristics of the rings to specific time periods. By comparing the sequence of rings in a wood sample with the reference dendrochronological curve for that particular species and geographical area, it is possible to accurately establish the period of growth and, consequently, the date of the wooden artefact. **Mensiocronology:** Mensiocronology is a direct and absolute dating method applied to brick artefacts or stone elements. It was developed in the 1970s by ISCUM and is based on the mathematical calculation of the dimensions of bricks or stone blocks belonging to a single building unit. In order to obtain a mensiocronological date, mensiocronological curves must already be available for the reference territory. These curves consist of databases that correlate the measurements of bricks or stone blocks with their date of manufacture or production, a certain date obtained from other sources (such as archival documents or scientific analyses dating the materials). Once the mensiochronological curves are available, simply enter the measurements of the elements you wish to date into the database to obtain a chronological estimate of their construction. [9,10,11].

How does it relate to BIM? One of the objectives of the research was to include all this data from the archaeological analysis of the architecture conducted on the study building within BIM; these data are essential for us to be able to achieve quality restoration. There was therefore a first application of the possibility of including detailed data, used in restoration projects, within the BIM; the case study analysed during the PRIN was that of the Albergo dei Poveri in Genoa. The Albergo dei Poveri is a 17th-century structure with several modifications in the following centuries. In this case, some parts of the building were studied using the methods of architectural archaeology, and particularly the method of brick mensiochronology. Here an attempt was made to make a comparison between manual and digital methods.

Emerging Research in Materials for Environment, and Civil Infrastructure - GeoME 5.5   Materials Research Forum LLC
Materials Research Proceedings 58 (2026) 108-115   https://doi.org/10.21741/9781644903933-15

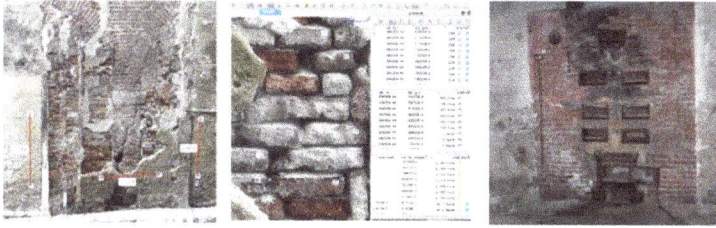

Fig.1. *1a-Vico Fate (Ge), The digital model created by Agisoft PhotoScan software is scaled inserting the distance between couples of coded targets placed on site; 1b- In case A and B, the bricks were identified on site with numbered labels to compare the results of direct and digital measurements block by block;1c- Tiled model view of case C, a wall of the ovens in the kitchen of the Albergo dei Poveri (source: [12]).*

In particular, the research conducted aimed to explore the application of photogrammetry to mensiochronology, a method of chronological dating based on the analysis of the dimensions and characteristics of bricks. Mensiochronology, as mentioned above, is a non-destructive method that uses statistical curves to correlate brick sizes with specific production periods (see fig. 1). This approach is particularly useful for understanding the construction history and transformations of historic buildings, as well as supporting conservation interventions. Photogrammetry offers significant advantages over direct measurement, including the reduction of on-site survey time, the possibility of off-site measurements and the involvement of multiple operators. The three case studies analysed were chosen because they featured brick walls from distinct historical periods: a historic house in Novi Ligure, a building in Genoa's vico Fate and a kitchen in the Albergo dei Poveri. The results show that photogrammetry is dependable for industrial bricks, but shows discrepancies for handmade bricks, influenced by the human factor. These initial conclusions indicate that photogrammetry can be useful for mensiochronology, but further research is needed to improve reliability, especially for complex surfaces and non-standardised bricks [13].

2.2 Further experimentation

The PRIN research project has intercepted the experimentation of BIM techniques on many important monuments: for example, to name but a few, the Duomo in Milan [13], the huge complex of the Albergo dei Poveri in Genoa [14]; the basilica of S. Maria di Collemaggio in L'Aquila, severely damaged by the 2009 earthquake [15], the Castel Masegra in Sondrio [16] (see fig.2 e fig.3). During these applications, which were conducted in constant comparison with real asset management processes, various solutions were evaluated, and all the activities involved in the cyclical management process of a listed building. In some cases, issues relating to the documentation of urban archaeological excavations were also included.

Emerging Research in Materials for Environment, and Civil Infrastructure - GeoME 5.5   Materials Research Forum LLC
Materials Research Proceedings 58 (2026) 108-115   https://doi.org/10.21741/9781644903933-15

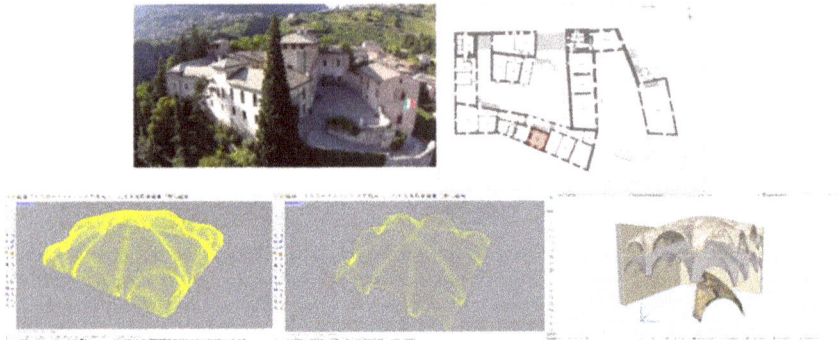

Fig.2. Masegra Castel reconstruction. Location of the umbrella vault. Reconstruction RHINO. Structure of constructive categories linked to a three-dimensional space of the three parts that make up the umbrella vault in the BIM environ (source: [17]

The use of BIM in the construction sector is now well established, and efforts are being made to extend its benefits to the cultural heritage sector in order to make processes more sustainable. However, the application of BIM to cultural heritage shows limitations in the representation of cultural value and in conservation activities. Research by Moyano et al. [17] highlights difficulties in interoperability between BIM software (FreeCAD, ArchiCAD and Revit) and in the connection between geometric and ontological models, as data is not automatically transferable. Some solutions to these difficulties aim to maximise efficiency and information retention through different approaches. For example, there are proposals for a single interoperable model or, alternatively, a platform that allows exchange between different specialised BIMs, coordinated by a common environment (CDE). The latter vision, based on interoperability and continuous coordination, may, in my opinion, represent a decisive step towards improving the management and quality of processes in the cultural heritage sector.

Conclusion
The research presented highlights the growing complexity and challenges in the field of heritage conservation and restoration, particularly in the Mediterranean context, where population growth, environmental pressures, and resource consumption pose significant threats to the built environment. Preserving historic buildings in a sustainable manner requires attention not only to traditional construction systems and unique architectural features but also to the careful integration of modern technologies. Today, heritage conservation faces two converging forces: the climate crisis and the digital transition. The climate crisis requires action in three steps: anticipate, adapt, and manage vulnerabilities. These vulnerabilities demand a sense of urgency and a recognition that heritage must be understood as a dynamic entity. We need to redefine not only its cultural and historical value but also its compatibilities with contemporary life and use. A cultural shift is needed in how we understand, manage, and preserve heritage. We must define rules, set visions, and make informed choices: what to conserve, how to conserve, and for whom. Competence is crucial; we must evaluate and assess the essential qualities of heritage. Questions such as what to record, for whom, for what purpose, and at what cost must guide the restoration process. Sustainability, both environmental and social, plays a key role. Environmentally, this may involve choosing low-impact materials, often those traditionally used, and intervening without disrupting a building's layout. Socially, cultural heritage serves as a testimony to cultural identity. Technology offers both opportunities and challenges. Recent advances, including BIM systems, photogrammetry, and methods from architectural archaeology, allow for more   accurate

Emerging Research in Materials for Environment, and Civil Infrastructure - GeoME 5.5   Materials Research Forum LLC
Materials Research Proceedings 58 (2026) 108-115   https://doi.org/10.21741/9781644903933-15

monitoring, documentation, and planning of interventions. Case studies, from the Albergo dei Poveri in Genoa to other national monuments, demonstrate both the potential and limitations of these technologies. In particular, integrating archaeological methods into BIM environments enables a more informed restoration process, but balancing model complexity, historical accuracy, and software interoperability remains challenging [18,19]. As Stefano della Torre noted during the Prin research, "We have to manage BIM before it manages us" [6] highlighting the need for careful and purposeful application of technology. While an abundance of data can aid restoration, it can also become a hindrance if not guided by critical historical judgment. Conservation should be seen as a strategic economic resource, an investment rather than a cost. Ultimately, sustainable heritage conservation depends on the degree of multidisciplinary collaboration that we will be able to develop in the future. Architects, engineers, materials scientists and digital technology experts will increasingly need to work together to develop tailored approaches that respect, for example, both the historical value and structural safety of buildings [20] and the environment. Future research should focus on improving the reliability of digital methods for irregular or non-standardised elements, enhancing BIM interoperability, and expanding databases such as those for chronotypological and mensiochronological analysis in architectural archaeology and their use in the BIM environment. Such integrated strategies could truly make heritage conservation more efficient, economical and respectful of cultural and historical authenticity, ensuring the preservation of historic buildings for future generations.

References

[1] Arayici Y. (2008) Towards Building Information Modelling for Existing Structures, "Structural Survey", 26.3: 210-222. https://doi.org/10.1108/02630800810887108

[2] Fai S., Graham K., Duckworth T., Wood N., Attar R., (2011) Building information modelling and heritage documentation. 18th International Conference on Virtual Systems and Multimedia. XXIII CIPA Symposium - Prague, Czech Republic.

[3] Volk, R., Stengel, J., Schultmann, F. (2014) Building Information Models (BIM) for existing buildings - literature review and future needs, Automation in Construction, 38, 2014:109-127. https://doi.org/10.1016/j.autcon.2013.10.023

[4] Logothetis S., Delinasiou A., Stylianidis E., (2015) Building Information Modelling for Cultural Heritage: a Review, ISPRS Annals of the Photogrammetry, Remote Sensing and Spatial Information Sciences, Volume II-5/W3, 2015 25th International CIPA Symposium 2015, 31 August - 04 September 2015, Taipei, Taiwan https://doi.org/10.5194/isprsannals-II-5-W3-177-2015

[5] Megahed, 2015: Megahed, N. A. (2015) Towards a Theoretical Framework for HBIM Approach in Historic Preservation and Management, "International Journal of Architectural Research", Volume 9 - Issue 3 - November 2015:130-147 https://doi.org/10.26687/archnet-ijar.v9i3.737

[6] Adami A., Della Torre S., Fregonese L., Mazzeri A., Scala B., Spezzoni A. (2017), Conoscenza e gestione del patrimonio costruito storico ("Studi sul recupero delle superfici decorate dell'architettura delle facciate della Cavallerizza e del Castello di San Giorgio in Palazzo Ducale di Mantova") le nuove frontiere del BIM, in Atti del convegno di Bressanone , ed Arcadia Ricerche, Venezia, pp.255-265.

[7] Pittaluga D. (2009), Questioni di archeologia dell'architettura e restauro, ECIG, Genova.

[8] Boato A., Pittaluga D. (2000), Building archaeology: a non-destructive archaeology in 15th world conference on non-destructive testing, Roma 2000, https://www.ndt.net/article/wcndt00/papers/idn365/idn365.htm, last accessed 2025/10/5.

[9] Mannoni, T., Milanese, M. (1988). Mensiocronologia. In: Archeologia e restauro deimonumenti. All'Insegna del Giglio, Firenze, pp. 383-402.

[10] Pittaluga, D., (2009), La mensiocronologia dei mattoni. Per datare, per conoscere e per comprendere le strutture storiche. ECIG, Genova

[11] Acacia S., Babbetto R., Casanova M., Macchioni E., Pittaluga D. (2017), Photogrammetry as a toolfor chronological dating of fired bricks structures in Genoa, The International Archives of the Photogrammetry, Remote Sensing and Spatial Information Sciences, Volume XLII-2/W5, 2017 26th International CIPA Symposium 2017, 28 August-01 September 2017, Ottawa, Canada , pp.749-753. https://doi.org/10.5194/isprs-archives-XLII-2-W5-749-2017

[12] Fassi F., Achille C., Mandelli A., Rechichi F., Parri S., (2015): A new idea ofBIM system for visualization, web sharing and using huge complex 3D models forfacility management. Int. Arch. Photogramm. Remote Sens. Spatial Inf. Sci., XL-5/W4, doi: 10.5194/isprsarchives-XL-5-W4-359-2015, pp. 359-366. la 2025/10/5. https://doi.org/10.5194/isprsarchives-XL-5-W4-359-2015

[13] Musso S.F., Franco G. (2014), The "Albergo dei Poveri" in Genova: conserving and using in the Incertainty and in the Provisional, in S. Della Torre, ed., ICT per il miglioramento del processo conservativo, Firenze, Nardini, 2014, pp. 41-50. ("UN BILANCIO DEL PROGETTO BHIMM - core.ac.uk")

[14] Oreni, D., Brumana, R., Della Torre, S., Banfi, F., Barazzetti, L., Previtali (2014), M., Survey turned into HBIM: the restoration and the work involved concerning the Basilica di Collemaggio after the earthquake (L'Aquila). ISPRS Annals of the Photogrammetry, Remote Sensing and Spatial Information Sciences, v.II, pp 267-273 https://doi.org/10.5194/isprsannals-II-5-267-2014

[15] Barazzetti L., Banfi F., Brumana R., Previtali M., Creation of parametric BIM objects from point clouds using nurbs, Photogramm. Rec. 30 (2015) 339-362. https://doi.org/10.1111/phor.12122

[16] Moyano J.,Pili A., Nieto-Juli J.E., Della Torre S., Bruno S. (2023), Semantic interoperability for cultural heritage conservation: Workflow from ontologies to a tool for managing and sharing data in Journal of Building Engineering, la 2025/10/5 https://www.sciencedirect.com/science/article/abs/pii/S2352710223021459?via%3Dihub.

[17] Scianna, A., Gaglio, G. F., & La Guardia, M. (2020). HBIM data management in historical and archaeological buildings. Archeologia e Calcolatori, 31(1), 231-252. Ursini, A. La metodologia BIM applicata ai beni culturali. Un caso specifico: il Sacro Monte di Varallo Sesia. Politecnico di Torino,2019. https://webthesis.biblio.polito.it/12043/1/tesi.pdf?utm_source=chatgpt.com la 2025/10/5.

[18] Carocci C. F., Borgia C., Costa M., Circo C., Indelicato D., Marino M., Lagomarsino S., Cattari S., Cianci F., Dal Bo A., Degli Abbati S., Ottonelli D., Romano C., Rossi M., Serafino N., Stagno G., Cifani G., Martinelli A., Castellucci A., Lemme A., Liris M., Martegiani F., Mazzariello A., Milano L., Morisi C., Petracca D., Tocci C., Pittaluga D., Vecchiattini R. 2010. Una metodologia per la conservazione dei centri storici danneggiati dal sisma: rilievo costruttivo e del danno, indagini ed indicazioni per il recupero di Casentino (AQ) [A methodology for the

Emerging Research in Materials for Environment, and Civil Infrastructure - GeoME 5.5   Materials Research Forum LLC
Materials Research Proceedings 58 (2026) 108-115   https://doi.org/10.21741/9781644903933-15

conservation of historic centres damaged by earthquakes: structural and damage surveys, investigations and recommendations for the restoration of Casentino (AQ)]. In Boscato G., Guerra F., Russo S., Sciarretta F., Sperotto E. (eds.) Sicurezza e conservazione nel recupero dei beni culturali colpiti da sisma. "Strategie e tecniche di ricostruzione ad un anno dal terremoto abruzzese." ("Pubblicazioni - Alberto Lemme") Volume degli atti del convegno, Venezia, 8-9 April 2010, pp. 282-291. Venezia: Università IUAV

[19] Pertica, S., (2002-2003). Architettura digitale. Tecnologie ed Applicazioni Informatiche per la Diagnostica. "Tesi di laurea, Facoltà di Architettura, Università degli Studi di Genova, Relatore: Torsello, B.P., Correlatori: Verri, A., Pittaluga, D., Odone, F., Genova"

Emerging Research in Materials for Environment, and Civil Infrastructure - GeoME 5.5   Materials Research Forum LLC
Materials Research Proceedings 58 (2026) 116-122   https://doi.org/10.21741/9781644903933-16

The urban void of the Mellah in Sefrou's Medina, reflections through integrated digital methodologies

Giovanni Pancani[1,*], Alberto Pettineo[1], Giovanni Minutoli[1] and
Houssem Dine Kouidhi[1]

[1]Department of Architecture, University of Florence, Florence, Italy

giovanni.pancani@unifi.it

Keywords: Photogrammetry, Digital Survey, Cultural Heritage, 3D Modelling, Conservation, Morocco

Abstract. The paper illustrates the outcomes of documentation activities carried out on the Mellah of Sefrou, an area historically subject to recurrent flooding, including a particularly significant event recorded in 1950 that caused severe damage to the built structures and led to an "urban void." In May 2017, within a workshop organised by the University of Florence in Sefrou, a digital survey campaign was conducted in the area affected by the floods, integrating direct measurements with Structure-from-Motion (SfM) photogrammetry. The aim was to produce three-dimensional datasets supporting the development of 3D models and two-dimensional drawings useful for describing the state of conservation of the structures. In parallel, an alternative workflow was tested, based exclusively on open-source data available online, and in this first phase, focused on modelling the topographic surface of the area. The investigation seeks to explore the potential of open-source geospatial data for generating three-dimensional models, assessing their reliability through comparison with field survey results, defining the effective scale of detail that can be achieved, and identifying possible application domains for products derived from such resources.

Introduction

The settlement of Sefrou, located in the eponymous province in the Fès-Meknès region at the foot of the Middle Atlas chain, has historically represented a centre of commercial and cultural exchange between the mountain areas and the fertile northern plains. During the twentieth century, the settlement faced serious flooding events, among which the most significant was in 1950, when the Oued Aggai River overflowed, causing widespread damage to the houses [1]. The steep morphology and a watershed geologically prone to erosion mean that the river continues to pose a potential hazard to the city due to the force of its floods [2-3]. The 1950 flood caused more than one hundred victims and devastated much of the Jewish quarter (Mellah), leading to its gradual abandonment. The space created was later transformed into a public square, which today serves as a memorial of that tragedy and an open-air laboratory for studying processes of decay, abandonment, and transformation in vulnerable historic contexts.

The absence of restoration works after the successive floods accelerated the area's decline, leaving buildings in ruin and a fragile heritage at risk of disappearance [4]. The fracture produced, both in the physical structure and in the cultural fabric, makes Sefrou an emblematic case of the challenges that the historic centres of the Maghreb are facing. Despite these transformations, the medina still today remains enclosed within its ancient walls, and its traditional souq continues to be lively, preserving artisan workshops and crafts that are now rare even in the most remote regions.

Emerging Research in Materials for Environment, and Civil Infrastructure - GeoME 5.5   Materials Research Forum LLC
Materials Research Proceedings 58 (2026) 116-122   https://doi.org/10.21741/9781644903933-16

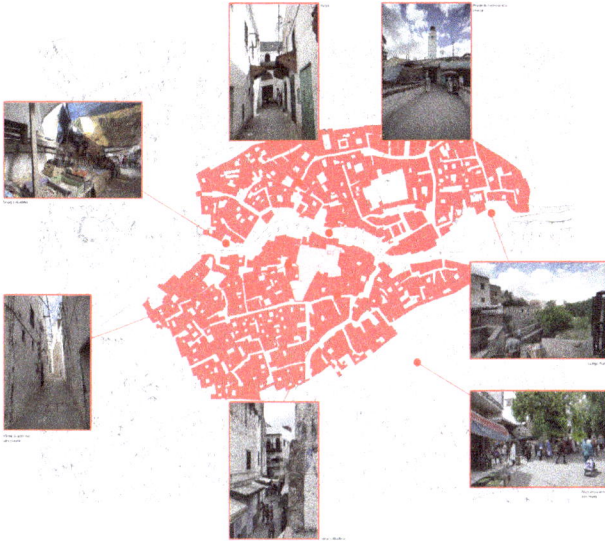

Fig. 1. Sefrou: map of the urban centre. The historic medina is highlighted in red, while the urban void of the Mellah is shown in pink. The Oued Aggai river bisects the city.

Considering cultural heritage in a broader sense, digital survey technologies have proven effective in documenting historic architecture in complex and at-risk contexts, contributing to creating models useful for analysing and conserving built heritage [5-7]. Specifically, the combined use of photogrammetric survey techniques and open-source data has shown the potential to obtain suitable models even in urban contexts that are difficult or impossible to access. Meanwhile, information modelling tools and semi-automated procedures based on GIS and open data jointly support historical reconstruction, digital management and the evolutionary interpretation of built environments [8-12]. GIS approaches have proven effective in analysing and classifying historic urban fabrics, as well as in the georeferenced management of urban morphology and planning purposes [13-14]. In this framework, remote digital techniques provide a concrete solution to logistical challenges in remote or unstable areas and offer a replicable methodological resource for documenting and enhancing fragile and stratified urban heritage.

Emerging Research in Materials for Environment, and Civil Infrastructure - GeoME 5.5   Materials Research Forum LLC
Materials Research Proceedings 58 (2026) 116-122   https://doi.org/10.21741/9781644903933-16

Fig. 2. Planimetric representation of the Mellah in Sefrou.

Direct Survey and SfM Photogrammetry

In May 2017, during an international seminar in the Fès-Meknès region, the research group from the University of Florence carried out a systematic survey campaign in the Mellah area to document the large "urban void" that had formed as a result of post-flood collapses in 1950.

The logistical situation, with the absence of terrestrial laser scanning equipment, required a flexible and integrated methodological approach based on a combination of direct surveying and photogrammetry techniques. This strategy enables the acquisition of detailed three-dimensional data and validates its reliability through comparison with manual control measurements [15-16].

For image acquisition intended for 3D reconstruction, high-resolution DSLR cameras were used (Sony Alpha 900 with Zeiss 24–70 mm lens), with shooting parameters set to maximise depth of field and minimise noise, thus ensuring the quality required for subsequent digital processing. Pre-processing of the images, including colour and exposure correction, helped to limit quality loss in the later processing stages. Photogrammetric processing was done using *RealityCapture* software, producing point clouds and textured mesh models. The reconstructions, at medium and high resolution for specific portions of the area, were functional for producing orthophoto plans at an architectural scale.

All images were calibrated and pre-processed according to standards consistent with the intended restitution scale (1:50, equal to 3 pixels/cm), considering inevitable quality losses and increasing the acquisition resolution by approximately 30%. The metric reliability of the photogrammetric outputs was validated through a direct survey using manual trilaterations and calibration targets placed on vertical and horizontal surfaces, ensuring geometric consistency across the different modelling levels.

From the photogrammetric model, orthoimages were also exported, providing a detailed and descriptive restitution of the architectural surfaces of the main facades. The graphic outputs, supported by an initial diagnostic analysis of the buildings facing the urban void, made it possible to document both traditional construction techniques and the main forms of decay.

Emerging Research in Materials for Environment, and Civil Infrastructure - GeoME 5.5   Materials Research Forum LLC
Materials Research Proceedings 58 (2026) 116-122   https://doi.org/10.21741/9781644903933-16

Open data and GIS for digital modelling

This phase aimed to evaluate the morphometric discrepancies between the model produced exclusively with open-source data and a certified survey baseline. The purpose was to consider the potential of open-source datasets as a viable tool for developing reliable 3D models. The 3D modelling process began with a preliminary analysis of the study area, seeking to obtain a three-dimensional representation of the terrain that was as consistent as possible with the real context, using the available topographic data. For this purpose, the online platform 'Contour Map Creator' was employed. This tool uses a script based on the Google Elevation API, Google Maps, jQuery, and the CONREC contouring algorithm [17-19], which allows contour lines to be generated from elevation data. Once the area of interest was selected, contour lines were exported at 0.10 m intervals in .KML (Keyhole Markup Language) format, compatible with platforms such as Google Earth and Google Maps.

The .KML file was then imported into QGIS. Within the software, the .KML was read as a set of vector layers, allowing for direct visualisation and interaction with the contour lines. Upon loading the file, the layer's attribute table was consulted, where the "description" column reported the elevation for each contour line in the text format "Contour ### m". To use this value numerically, it was necessary to extract and insert the elevation into a new column labelled "Elevation". This operation was carried out using the QGIS field calculator with the following expression: substr("description", strpos("description", ' ') + 1, strpos("description", 'm') strpos("description", ' ') - 2)

This expression enabled the numerical elevation value to be isolated and interpreted by QGIS as a Z-value. After this phase, the main urban features were vectorised. Using the "QuickOSM" plugin integrated in QGIS, available geographic data from OpenStreetMap for the Sefrou area - including roads, watercourses and simplified buildings - were downloaded and imported as separate layers. The "Qgis2threejs" plugin was used to create the three-dimensional model, allowing 3D visualisations to be generated directly in QGIS and exported in formats compatible with other modelling software.

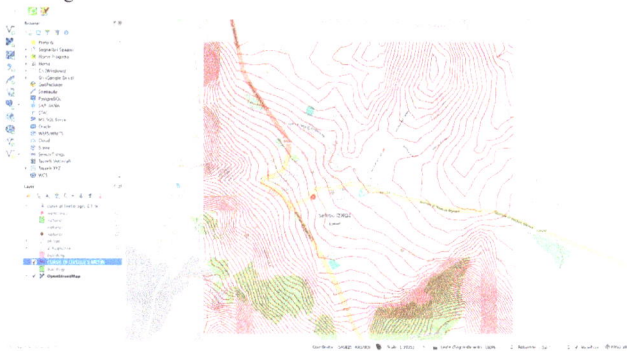

Fig. 2. Assessment of the geolocation accuracy for data imported from Contour Map Creator and Open-StreetMap.

The contour line layer was selected, assigning the "Z-value" property to the data in the "Elevation" column. The entire model was exported in .GLTF (GL Transmission Format), suitable for import into modelling environments and NURBS-based software such as McNeel Rhinoceros 3D.

After importing the file, scales and measurements were checked to ensure consistency with the units used in QGIS. Based on these data, a terrain mesh was generated, providing a three-

Emerging Research in Materials for Environment, and Civil Infrastructure - GeoME 5.5   Materials Research Forum LLC
Materials Research Proceedings 58 (2026) 116-122   https://doi.org/10.21741/9781644903933-16

dimensional representation of ground morphology corresponding to the city's urban layout. The polylines for roads and watercourses were projected onto the mesh, maintaining topological coherence with the elevation model.

Comparative analysis and future directions

After the modelling process, the focus was placed not so much on the comparative analysis of possible structural changes between the two reconstruction methods - a result that, in the absence of a direct site visit, would risk remaining purely hypothetical - but rather on the possibility of validating the quality and reliability of the morphological representations obtained from open-source data. Therefore, the main objective of this phase was to assess whether, and to what extent, a 3D model of the terrain produced entirely remotely can provide a sufficiently accurate picture compared to data collected on site with traditional and instrumental survey methods.

Fig. 4. *Comparative analysis between the 2D drawing produced in 2017 and the ground section line generated from the 2025 model (in red).*

For this analysis, the topographic sections exported from the digital model were traced along the same axes used in the survey data. The resulting profiles were then superimposed and compared point by point to assess the deviation between the elevations returned by the two systems. In total, 106 sections distributed along the central axis of the study area were considered, allowing the accuracy of the open-source model to be verified against the reference data. The recorded values show a deviation ranging from -0.316 m to +0.062 m. Although this margin of error may be significant in specific contexts, remote modelling nevertheless offers a useful representation of Mellah urban morphology. The comparative analysis of the topographic surface allowed for several methodological and operational considerations. The comparison between the sections derived from open-source data and those obtained through direct surveys showed a generally consistent and sufficiently accurate representation of the terrain morphology. The deviations remained within acceptable margins for preliminary analysis or for representation purposes aimed at spatial understanding of the site and/or initiating actions for its enhancement.

Competing Interests
The authors declare no conflicts of interest relevant to the content of this paper.

References

[1] Idrissi BE, Cherai B, Hinaje S, Mehdi K (2018) Climatic variability and its influence on water resources in the northern part of the Middle Atlas Moroccan. LARHYSS (39): 155-179.

[2] Imaouen A, Gourari L, Labraimi M, Essahlaoui A, Boukil A (2023) Characterization of physical, hydrological, and geological features of Oued Aggay watershed upstream of Sefrou City using GIS and remote sensing approaches. Arab J Geosci 16 https://doi.org/10.1007/s12517-023-11577-w

[3] Lahsaini M, Tabyaoui H (2018) Modélisation hydraulique mono dimensionnel par HEC-RAS, application sur l'Oued Aggay (Ville de Sefrou). Eur Sci J 14(18):110 https://doi.org/10.19044/esj.2018.v14n18p110

[4] Dipasquale L (2021) La medina di Sefrou: comprendere la cultura costruttiva. In: Hadda L (a cura di) Médina. Espace de la Méditerranée. Firenze University Press, pp 43-54

[5] Bertocci S, Arrighetti A, Bigongiari M (2019) Digital survey for the archaeological analysis and the enhancement of Gropina archaeological site. Heritage 2(1):848-857 https://doi.org/10.3390/heritage2010056

[6] Galassi S, Bigongiari M, Tempesta G, Rovero L, Fazzi E, Azil C, Pancani G (2022) Digital survey and structural investigation on the triumphal arch of Caracalla in the archaeological site of Volubilis in Morocco. Int J Archit Herit 16(6):940-955 https://doi.org/10.1080/15583058.2022.2045387

[7] Parrinello S (2024) La permanenza del segno in un linguaggio senza contorni e in continua evoluzione. TRIBELON 1:4-7 https://doi.org/10.36253/tribelon-2849

[8] Meyer E, Grussenmeyer P, Perrin J (2007) Virtual research environment for the management and the visualization of complex archaeological sites. 7th Int Symp Virtual Reality, Archaeol Cult Herit: 1-8.

[9] Gorreja A, Di Stefano F, Piccinini F, Pierdicca R, Malinverni ES (2021) 3D GIS for a smart management system applied to historical villages damaged by earthquake. In: ARQUEOLÓGICA 2.0 & GEORES, pp 255-260. https://doi.org/10.4995/arqueologica9.2021.12132

[10] Pettineo A, Dell'Amico A, Picchio F, Parrinello S (2024). H-BIM e GIS per l'analisi e la ricostruzione filologica del castello di Almencir in Spagna. DN, 14, 6-16.

[11] La Placa S, Parrinello S (2024) Scan to GIS methodologies for the multi-scalar representation of water landscapes. In: EMUNI International Conference.

[12] Giannopoulou M, Vavatsikos AP, Lykostratis K, Roukouni A (2014) Using GIS to record and analyse historical urban areas. TeMA J Land Use Mobil Environ

[13] Baratin L, Bertozzi S, Moretti E (2015) GIS intelligence for a cutting-edge management of 3D cities. In: 2015 Digital Heritage, vol 2, pp 93-96. IEEE https://doi.org/10.1109/DigitalHeritage.2015.7419460

Emerging Research in Materials for Environment, and Civil Infrastructure - GeoME 5.5   Materials Research Forum LLC
Materials Research Proceedings 58 (2026) 116-122   https://doi.org/10.21741/9781644903933-16

[14] Pancani G, Gentili M (2018) Urban survey of the Mellah of Sefrou (Mo). Dialogues Cult Herit:555-558

[15] Pancani G (2019) Sefrou, il rilievo del vuoto urbano della mellah prodottosi in seguito alle alluvioni del Oued Aggai. In: Patrimonio in divenire, conoscere, valorizzare, abitare, pp 917-928. Cangemi Editore

[16] Bourke PD (1987) A contouring subroutine. Byte 12(6):143-150

[17] Maunder CJ (1999) An automated method for constructing contour-based digital elevation models. Water Resour Res 35(12):3931-3940 https://doi.org/10.1029/1999WR900166

Emerging Research in Materials for Environment, and Civil Infrastructure - GeoME 5.5   Materials Research Forum LLC
Materials Research Proceedings 58 (2026) 123-129   https://doi.org/10.21741/9781644903933-17

Fast-survey techniques for digital mapping and understanding the Ksar of Ait Ben Haddou

Alberto Pettineo[1,*] and Giovanni Pancani[1]

[1]Department of Architecture, University of Florence

alberto.pettineo@unifi.it

Keywords: Digital Survey, Photogrammetry, Earthen Architecture, Digital Heritage, Morocco

Abstract. This study examines the Ksar of Ait Ben Haddou in southern Morocco to evaluate the potential of rapid and non-invasive techniques for documenting and analysing complex vernacular heritage. The workflow combined videogrammetry from unmanned aerial systems (UAS) with terrestrial laser scanning, prioritising speed, portability, and minimal equipment while ensuring metric reliability. Aerial, radial, and nadir video paths were used to capture the site's volumetric complexity and the extracted frames were processed through Structure from Motion–Multi View Stereo (SfM-MVS) algorithms to generate photogrammetric dense point clouds. In two key areas, the fortified granary on the hilltop and the noble zone within the lower settlement, a high-resolution TLS point cloud provided accurate metric references, enabling the calibration and validation of videogrammetric data. The resulting point cloud delivers a detailed representation of the ksar's external morphology, offering visualisation and qualitative support for architectural interpretation. Comparative analyses between UAS-based and laser-scanned datasets highlight the degree of dimensional accuracy achievable through rapid videogrammetry, as well as its capacity to capture building elements relevant for morphological reading. This assessment extends to the potential use of such datasets as foundations for interpretative or parametric modelling workflows, ensuring reliable representations from survey data. The research highlights the potential of fast-survey methodologies for preliminary analysis and documentation and as a starting point for interpretative processes concerning architectural form, construction logic, and typological variation - all enhanced by the visualisation and comprehension opportunities of structuring 3D information models.

Introduction

The Ksar of Ait Ben Haddou is one of southern Morocco's most emblematic examples of earthen architecture, recognised as a UNESCO World Heritage Site since 1987. Its strategic position along ancient caravan routes and integration within the pre-Saharan cultural landscape underlines its architectural and historical importance [1].

The site, consisting of multi-storey adobe buildings enclosed by fortified walls, reflects centuries of socio-cultural evolution shaped by trans-Saharan trade, Berber artisanal craftsmanship, and the principles of earthen construction. The social and residential structure of the site has undergone profound transformations, marked by depopulation and a growing musealisation that has progressively emptied the Ksar of its residential function. Today, only a few buildings remain inhabited, while the rest of the village is perceived mainly as a scenographic and touristic space. This reflects a broader crisis of habitability affecting ksour and kasbahs across southern Morocco, where patrimonialisation practices, though aimed at preservation, often detach these places from their original communities and uses [2][3]. Depopulation, in favour of new settlements, also entails the loss of traditional building and maintenance techniques, increasingly replaced by modern construction technologies that threaten the architectural identity of these landscapes. Addressing the challenges of preserving the region's cultural heritage thus requires survey strategies capable of capturing and representing the architectural and social specificities of the ksour at multiple scales.

Fig.1 *Ait Ben Haddou complex captured by UAV, general view of the site and its landscape context.*

In such scenarios, acquisition time limitations and the portability of instruments become key operational challenges, especially in remote areas or situations where accessibility is restricted. These conditions have required developing and adopting fast-survey strategies that combine documentary effectiveness, rapid execution, and metric accuracy. Using swift survey methodologies - combining videogrammetry from UAV footage and terrestrial laser scanning (TLS) - emerges as an effective alternative for generating point clouds that are useful for the morphological analysis of the site and also for structuring accurate and interoperable 3D models.

State of the Art

In recent years, techniques such as terrestrial laser scanning (TLS) and photogrammetry based on SfM-MVS algorithms - which reconstruct 3D models through the automated processing of images acquired from multiple viewpoints - have made it possible to produce accurate and highly descriptive spatial representations [4].

Alongside these consolidated approaches, recent experiments demonstrate that videogrammetry can achieve suitable levels of accuracy even with lightweight, non-professional equipment, making it particularly effective in operational contexts where efficiency, speed, and simplicity are priorities, especially in complex and extensive environments [5-7]. This evolution has contributed significantly to the widespread adoption of integrated approaches that, starting from generating point clouds and 3D models, now enable the development of databases, information systems, and digital twins to support the documentation and management of heritage sites [8-10]. Despite advances in video tools and the increasing accessibility of technologies, videogrammetry remains less explored than traditional photogrammetry, especially in academic and professional contexts. Considering the limited systematic applications of videogrammetry, this study seeks to contribute to the methodological discussion by offering a comparative reflection on techniques and tools [11].

Applied Methodology

The Ksar of Ait Ben Haddou presents a morphologically complex context, where survey planning must consider accessibility constraints, time limitations, and the specific objectives of the documentation campaign [12]. Data acquisition was organised into two main stages: aerial survey using a UAV drone and terrestrial laser scanning survey to develop a workflow ensuring minimal equipment, optimising both spatial coverage and the metric accuracy of the collected information.

Emerging Research in Materials for Environment, and Civil Infrastructure - GeoME 5.5   Materials Research Forum LLC
Materials Research Proceedings 58 (2026) 123-129   https://doi.org/10.21741/9781644903933-17

The aerial survey was carried out using a DJI Mavic 3 Enterprise1. A total of 10 videos, each with an average duration of about three minutes, were captured along radial and nadiral trajectories, documenting both architectural details and the broader settlement and landscape context. Flight planning prioritised operational flexibility, with variations in altitude and camera orientation to optimise coverage of the site's vertical and horizontal surfaces [13]. For the definition and metric validation of the dataset, terrestrial laser scanning (TLS) was conducted using a Leica BLK3602. The scans were concentrated in two significant areas: the Agadir (fortified granary) at the top of the Ksar and a portion of the settlement corresponding to the noble part of the village, situated in the lower section. To ensure spatial continuity and proper integration between the surveyed areas, the main connecting route between the summit and the noble areas of the Ksar was also mapped.

The decision to implement the TLS survey on these targeted portions, interconnected by the main circulation axis, enabled the definition of a polygonal scan path spanning the outermost limits of the Ksar. This methodological choice provided a longitudinal control network, crucial for the subsequent scaling and metric validation of the video-based model and ensuring geometric consistency across the different survey segments3. The TLS data were processed in Leica Cyclone Core, aligning the scans through a cloud-to-cloud registration workflow.

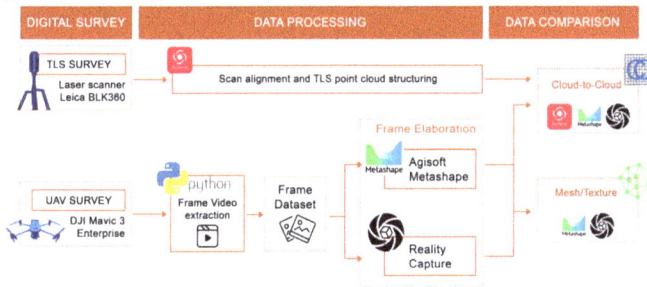

Fig. 2 *Applied methodology*

Drone videos were sampled regularly to extract frames suitable for 3D reconstruction using SfM-MVS techniques. Since the GPS data were not embedded in the metadata of the frames but were instead recorded separately in an .SRT (subtitle track) file, it was necessary to carry out an extraction and association process for geolocation data. The .SRT file, automatically generated by the drone's onboard software (DJI), contains a series of time blocks with flight information corresponding to each video frame, including latitude, longitude, and absolute altitude. These data were extracted using a custom Python script in a Windows environment. The script performed the following operations: (1) Extraction of frames from videos at defined regular intervals using OpenCV library; (2) Parsing the .SRT file generated by the drone to read and interpret positional data (Lat., Lon., Alt.) associated with each frame. (3) Creation of a .CSV file for each frame, containing the image name, GPS data, timestamp, and main camera settings, formatted for compatibility with image-based modelling software. (4) Writing geolocation data directly into the EXIF metadata of the exported images using the Piexif library.

---

1 The DJI Mavic 3 Enterprise is equipped with a 4/3" CMOS sensor (20 MP), a 24 mm equivalent lens, aperture from f/2.8 to f/11 and supports video recording up to 4K (3840×2160) at 30 fps.
2 The Leica BLK360 is a terrestrial time-of-flight laser scanner with a maximum range of 60 metres and an accuracy of 4 mm at 10 metres. It can acquire point clouds at up to 360,000 points per second, with a horizontal/vertical field of view of 360°/300°.
3 A total of 20 scans were performed in the lower part of the village, 15 in the upper part, and 30 along the main connecting route, all acquired in colourless mode to optimise survey times. The TLS acquisition was aimed exclusively at metric control and validation of the scale of the image-based models, rather than at generating a detailed coloured model of the complex.

Emerging Research in Materials for Environment, and Civil Infrastructure - GeoME 5.5   Materials Research Forum LLC
Materials Research Proceedings 58 (2026) 123-129   https://doi.org/10.21741/9781644903933-17

Writing metadata directly via a Python script rather than relying on standard photogrammetric software tools allows for managing the entire pipeline in a flexible and customisable way. This method does not bind the dataset to the use of any specific software and thus maintains the complete portability of the geotagged images.

Fig. 3 Workflow of the frame's alignment process and the resulting model of the entire complex processed in RealityCapture software.

The result of this process is a dataset of georeferenced frames4, ready to be imported into 3D reconstruction software and usable in any image-based processing environment. To evaluate the impact of software choices on the quality and reliability of reconstructed data, the georeferenced images were processed in parallel in two photogrammetric environments: Agisoft Metashape (MTH) and RealityCapture (RC). This approach allowed for comparison of the results, identifying the workflow best suited to the case study's specific requirements. Each software package processed the same georeferenced frame dataset for objective comparison. Metashape's workflow included importing, aligning, and generating the photogrammetric point cloud, which was then scaled and aligned to the TLS data (local UCS). In RealityCapture, the same procedure was applied, using identical reference points to ensure consistency in the comparison. GPS coordinates facilitated the initial alignment, which was later converted into the local UCS for scaling based on TLS points. Finally, the system was reconverted to the original geographic coordinates, providing a database that can be used and integrated for territorial-scale applications.

Results and Future Directions

The resulting point clouds from the two frame elaboration workflows were compared metrically and qualitatively to identify the most effective solution for representing and analysing the Ksar. The assessment of metric reliability was conducted on the point cloud using a cloud-to-cloud (C2C) analysis in *CloudCompare*, limited to a sample area in the lower portion of the Ksar (one of the Tighremt structures in the Noble area), where TLS data were also available as reference. This allowed for homogeneous comparison conditions and an objective evaluation of the metric and morphological performance of the two workflows.

---

4 A total of 1,410 frames were extracted from the video, each at a resolution of 3840×2160 pixels (4K UHD) and 96 DPI.

Emerging Research in Materials for Environment, and Civil Infrastructure - GeoME 5.5   Materials Research Forum LLC
Materials Research Proceedings 58 (2026) 123-129   https://doi.org/10.21741/9781644903933-17

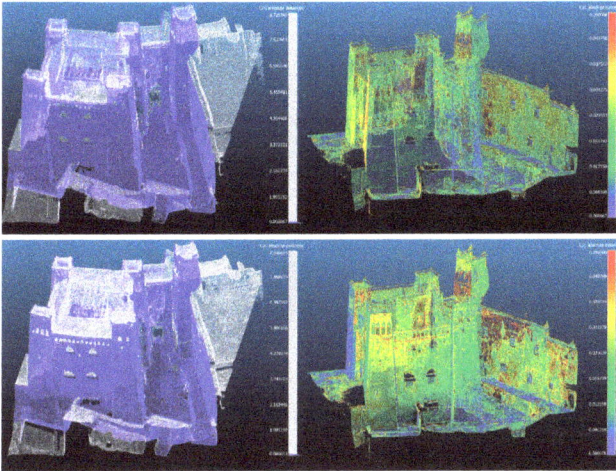

Fig. 4 C2C deviation maps for Agisoft Metashape (Fig. 4a) and RealityCapture (Fig. 4b) on a portion of the complex (Tighremt): (left) view with outlier points active (in white, the roofs and areas present only in drone data); (right) filtered view ≤ 0.05 m showing only points spatially corresponding to the TLS cloud

The Cloud-to-Cloud Distance function (C2C Distance) was used to calculate spatial deviations between the video-derived point clouds and the TLS model. Results were visualised using chromatic distance maps and analysed statistically (mean values, standard deviation, and error distribution). The analysis of scalar maps (Fig. 4) confirms a substantial parity in metric accuracy between the two software, with mean deviation from TLS consistently below 5 cm, in line with expectations for surveys at urban and territorial scales. The RC point cloud showed an average deviation of about 4 cm from the TLS model, while MTH's was about 3 cm. Although the latter is marginally more accurate in metric terms, the RC model provides a point cloud with a higher point density and geometric quality in representing architectural forms.

This experience highlights the effectiveness of the applied fast-survey workflow and its potential application to further cases of earthen architecture in Morocco's pre-Saharan areas. The frame-based point cloud, eventually integrated with TLS data, of the entire Ksar, now provides a foundation for developing descriptive systems of the architectural and morphological features of the complex, opening up new prospects for analysis, thematic representations, and digital enhancement of the site. The point cloud is a useful resource for structuring 3D information models, serving as a basis for reverse, parametric, and analytical modelling workflows, and providing a geometric foundation for online platforms, virtual environments, and interactive systems for cultural heritage management and dissemination [14–16]. The development of such information models constitutes support for management, maintenance, and conscious conservation strategies, as well as for communication and digital enhancement initiatives that are in line with the most recent perspectives of Heritage Science.

Competing Interests
The authors declare no conflicts of interest relevant to the content of this paper.

Acknowledgments
The documentation activities were conducted during an international workshop held at Tamnougalt and Ait Ben Haddou, jointly coordinated by the Dept. of Geology, Semlalia Faculty of Science, Cadi Ayyad University (UCA), and the Dept. of Architecture, University of Florence (UNIFI). The workshop was organised by Professors Mounsif Ibnoussina (UCA), Sandro Parrinello (UNIFI), and Giovanni Pancani (UNIFI), with the collaboration of Professors Admou Khaoula, El Abbassi Fatima Zahra, Bahammou Younes, Berroug Fatiha (UCA), and Letizia Di Pasquale, Lamia Hadda (UNIFI), as well as Fabio Fratini (CNR Florence), Alberto Pettineo (UNIFI), and Dr. Monica Lusoli (UNIFI), and actively involved students from the Department of Architecture, UNIFI, and Cadi Ayyad University (UCA).

Fig. 5 3D Digital database with TLS and frame-based data elaborated with RealityCapture.

References

[1] Werner, L. (1993). Ait Ben Haddou, a desert-born model for urban design. The UNESCO Courier: A Window Open on the World, 46(6), 46-47.

[2] Dłużewska A, Dłużewski M (2017) Tourism versus the transformation of ksours - Southern Morocco case study. Bull Geogr Socio-Econ Ser 36:77-86 https://doi.org/10.1515/bog-2017-0015

[3] Abyaa Z, El Harrouni K, Degron R, Aljem S (2025) From disintegration to valorisation: what perspectives and multi-scale approaches towards a sustainable territorial design for the architecture of Kasbahs and Ksour in southern Morocco? In: Correia M, Costa E, Roque J (eds) Vernacular architecture. Materials Research Forum LLC, pp 294-301 https://doi.org/10.21741/9781644903391-34

[4] Wojciechowska G, Łuczak J (2018) Use of close-range photogrammetry and UAV in documentation of architecture monuments. E3S Web Conf 71:00017 https://doi.org/10.1051/e3sconf/20187100017

[5] Torresani A, Remondino F (2019) Videogrammetry vs photogrammetry for heritage 3D reconstruction. ISPRS Arch 42 https://doi.org/10.5194/isprs-archives-XLII-2-W15-1157-2019

[6] Musicco A, Rossi N, Verdoscia C (2023) Accuracy evaluation of smartphone-based videogrammetry for cultural heritage documentation process. ISPRS Arch 48. https://doi.org/10.5194/isprs-archives-XLVIII-M-2-2023-1119-2023

[7] Pettineo A (2022) Videogrammetry for the virtual philological reconstruction of the Scaliger fortifications in the territory of Verona. In: D-SITE, pp 104-111. Pavia University Press.

[8] Parrinello S, Picchio F (2023) Digital strategies to enhance cultural heritage routes: from integrated survey to digital twins of different European architectural scenarios. Drones 7(9):576 https://doi.org/10.3390/drones7090576

[9] Parrinello S, Porcheddu G (2023) Documentation procedures for rescue archaeology through information systems and 3D databases. In: Beyond digital representation, pp 761-778. Springer Nature Switzerland https://doi.org/10.1007/978-3-031-36155-5_49

[10] Dell'Amico A (2020) The application of fast survey technologies for urban surveying: the documentation of the historic center of Santa Cruz de Mompox. In: D-SITE, pp 132-141. Pavia University Press

[11] Ortiz-Coder P, Cabecera R (2021) Accurate 3D reconstruction using a videogrammetric device for heritage scenarios. ISPRS Arch 46(M-1):499-506 https://doi.org/10.5194/isprs-archives-XLVI-M-1-2021-499-2021

[12] Picchio F, Pettineo A (2019) Fotogrammetria per la creazione di banche dati utili alla lettura e alla comprensione dei sistemi fortificati. In: Dalmazia e Montenegro, le fortificazioni venete nel bacino del Mediterraneo orientale. Pavia University Press, Pavia, pp 89-96

[13] Parrinello S, Picchio F (2019) Integration and comparison of close-range SfM methodologies for the analysis and the development of the historical city center of Bethlehem. ISPRS Arch 42:589-595 https://doi.org/10.5194/isprs-archives-XLII-2-W9-589-2019

[14] Soler, F., Melero, F. J., & Luzón, M. V. (2017). A complete 3D information system for cultural heritage documentation. Journal of Cultural Heritage, 23, 49-57. https://doi.org/10.1016/j.culher.2016.09.008

[15] Parrinello S, Pettineo A (2025) Databases and information models for semantic and evolutionary analysis in fortified cultural heritage. Heritage 8(1):29 https://doi.org/10.3390/heritage8010029

[16] Parrinello S, Dell'Amico A, Galasso F, Porcheddu G (2024) Virtual spaces for knowledge preservation: digitization of a vanished archaeological excavation. In: Advances in representation: new AI- and XR-driven transdisciplinarity, pp 237-254 https://doi.org/10.1007/978-3-031-62963-1_14

Emerging Research in Materials for Environment, and Civil Infrastructure - GeoME 5.5   Materials Research Forum LLC
Materials Research Proceedings 58 (2026) 130-138   https://doi.org/10.21741/9781644903933-18

Thermal efficiency of traditional building materials in diverse climates: The literature review

Doha CHBARI[1,a]*, Oumaima Ait Rami[1,b] and Kaoutar Ouali[1,c]

[1]Materials and Nanomaterials for Photovoltaics and Energy Storage Laboratory (MANAPSE Lab), Faculty of Sciences, Mohammed 5 University in Rabat, Rabat, Morocco

[a]dchbari@gmail.com, [b]oumaima_aitrami@um5.ac.ma, [c]k.ouali@um5r.ac.ma

Keywords: Thermal Performance, Traditional Building Materials, Energy Efficiency, Climatic Conditions, Sustainable Construction

Abstract. This paper evaluates the thermal characteristics of traditional building materials commonly used in five countries with diverse climatic conditions: Morocco, Fin-land, India, Mexico, and Japan. Traditional materials such as rammed earth, stone, date palm fiber, wood, lime, adobe, bamboo, and thatched roofs were evaluated in terms of their thermal conductivity and their capacity to moderate indoor temperatures effectively. Using literature review methodology, comparative studies were conducted to quantify decreases in cooling and heating loads, indoor thermal comfort enhancements, and possible energy savings for each material. Principal results indicated that Moroccan rammed earth significantly re-duces the heating demand by up to 32%, Finnish wood construction contains ap-proximately 69% less embodied energy compared to concrete, Indian earthen plasters reduce indoor temperatures by a significant amount ranging from 4°C to 6°C, Mexican adobe walls has a thermal conductivity of 0.69 W/m·K effectively reducing the indoor temperatures significantly, and Japanese thatched roofs re-duce interior temperatures by about 5°C to 7°C compared to the exterior temperatures. The results highlight the importance of employing local low-energy traditional materials in significantly enhancing thermal comfort, energy efficiency, and environmental sustainability, thereby qualifying them as materials of high interest for sustainable building practice particularly under various climatic conditions.

Introduction

Historically, societies used locally adapted materials such as earth, wood, stone, and lime, while the Industrial Revolution introduced reinforced concrete, steel, and glass, thus enabling faster construction that was standardized in response to demands for stronger and quicker building.

Traditional materials possess several advantages over modern ones, including low-energy production, non-pollution, environmental friendliness, and recyclable nature within natural and economic cycles.[1], The earliest building techniques relied on natural materials such as stone, wood, earth, and bricks. Their abundance and versatility made them the cornerstone of traditional architecture.[2], In the past few years, new materials including steel and cement have largely superseded traditional indigenous building materials based on their greater lifespan, lower maintenance needs, resistance to rust and rotting, and readiness with which they facilitate the process of construction. However modern building materials exhibit high energy consumption and environmental detriment [3].

This paper is a literature review that evaluates multiple studies regarding the thermal efficiency of traditional building materials across five nations with di-verse climatic conditions: Morocco (Mediterranean/arid), Finland (subarctic), India (tropical/monsoon), Mexico (desert/tropical varied), and Japan (temperate with seasonal variations). → (Fig. 1)

Emerging Research in Materials for Environment, and Civil Infrastructure - GeoME 5.5   Materials Research Forum LLC
Materials Research Proceedings 58 (2026) 130-138   https://doi.org/10.21741/9781644903933-18

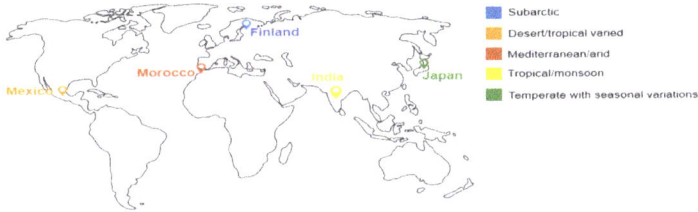

Fig. 1 *Countries selected for the comparative study.*

Methods

This literature review was conducted to assess the thermal performance of traditional building materials in five countries which are representative of various climatic zones. The research process involved several key steps:

- Literature search using keywords such as "traditional building materials," "vernacular architecture," and combinations such as "wood in Japan" or "rammed earth in Morocco."
- Additional literature searches were conducted for all identified materials to determine their thermal performance and energy contribution to efficiency.
- Information gathered included peer-reviewed articles, case studies and simulation studies.
- Parameters like thermal conductivity, thermal resistance, energy savings, and temperature.
- Results were then presented by country and material.

Results and Discussion

Morocco

The climate varies in Morocco from Mediterranean to arid desert: the coastal areas have mild, wet winters and hot, dry summers; 2023 mean temperatures exceeded 22°C in the south and 12–18°C in mountains with 18–20°C elsewhere, while summer peaks in the interior reached over 40°C, and winter nights in mountainous regions fell below 0°C [4], In response to these climatic variations, traditional Moroccan buildings have adapted by using local materials such as raw earth, stone, rammed earth, and lime. Several papers have discussed traditional construction materials:

Rammed earth

Rammed earth can be defined as a building material derived from compacting natural soil layers, normally in the form of a mixture of pebbles, clay, silt, sand, gravel, and occasionally pebbles, between temporary formworks to create solid structural walls.[5]

Rammed earth, which is a major part of Morocco's architectural heritage, has recently regained interest due to its energy, environmental, socio-economic, and aesthetic advantages. However, its thermal and ecological performance remains poorly studied. [6]. The potential of rammed earth as an energy-efficient solution is assessed in [7] by comparing the thermal behaviour of rammed earth with concrete buildings using dynamic simulations. The findings indicate that rammed earth construction performs better in terms of operative and radiant temperatures, minimizing the feeling of "cold walls" due to differences between indoor air and surface temperatures.

There is also a considerable decrease in heat losses through the walls, top and bottom floors, compared to masonry construction. The study additionally indicates that heating requirements within the rammed earth building are reduced by up to 32% compared to the masonry alternative. → Table 1 shows the primary findings of this study:

Table 1. *Summary of simulation results. [7]*

Option	Radiant temperature (°C)	Operative temperature (°C)	Heat loss in Walls (KW)	Heat loss in Roofs (KW)	Heating demand [kWh·m²]	Cooling demand [kWh·m²]
Masonry	15,12	17,54	7,75	2,73	49,94	507,32
Rammed earth	16,65	18,3	4,11	1,94	33,96	668,68

Date palm fibre

The study conducted by M. Belhous in Effect of a Material Based on Date Palm Fibers on the Thermal Behavior of a Residential Building in the Atlantic Climate of Morocco, evaluates the impact of date palm fiber insulation on the thermal efficiency of a building in an Atlantic climate, showing that this material significantly improves indoor comfort by reducing heating and cooling loads, lowering energy demand by up to 25% for cooling and 18% for heating, while also cutting costs and CO2 emissions. [10]

the study of M. Boumhaout shows that the addition of date-palm fibers enhances the thermal performance of mortar considerably: thermal conductivity decreases by up to 70% (0.795-0.243 W/m·K), while diffusivity and effusivity are reduced for more than 50%, by reducing heat transfer and improving indoor temperature stability-heating and cooling loads could be lower by means of a lightweight, sustainable insulating material. [11]

Stone

Another study conducted by N.Eraza [12] explores the application of confined stonewall structures (MPC) to meet thermal comfort and energy savings in Moroccan housing in six climatic zones. The findings show that MPC walls with a thermal resistance of 0.62 m².K/W reduce heating demands due to their high thermal inertia, keeping the indoor temperature under 30°C in summer and over 15°C in winter without mechanical cooling or heating. The superimposition of a 7 cm layer of superficial soil and fiber insulation reduces the U-value to 0.62 W/m².K, which is acceptable according to Moroccan thermal regulations (RTCM). Indoor humidity levels are also stable, ranging between 40% to 80%. MPC buildings, on the whole, represent a cost-saving and green solution for rural homes, reducing both CO_2 emissions and energy use.

Finland

Finland has a subarctic climate throughout most of the nation, temperate continental in the south. The climate is characterized by harsh seasonal extremes. The winters are long, cold, and snowy, with temperatures often below 0°C, dropping to -6°C in northern regions in December. Summer is brief but quite moderate, particularly in the south of the country, with average summer temperatures of between 16°C and 18°C in July and August. [13]

Wood

Wood is a natural building material that has traditionally been used in Finland, primarily in the form of timber or logs. The material derives from the country's rich forest resources and is formed into huge horizontal pieces, especially load-bearing walls. [13]

Wood buildings in Finland are highly energy-efficient due to the intrinsic insulating capacity of wood and its low embodied energy. The thermal conductivity of softwood is 0.12-0.14 W/m·K, which is good for insulation. Wood-frame and mass timber construction can reduce embodied energy by up to 69% compared to concrete construction. Cross-laminated timber (CLT) construction conserves up to 38% of the annual heating energy, and hybrid timber systems minimize thermal bridging and improve airtightness. [14]

Emerging Research in Materials for Environment, and Civil Infrastructure - GeoME 5.5   Materials Research Forum LLC
Materials Research Proceedings 58 (2026) 130-138   https://doi.org/10.21741/9781644903933-18

According to a study that involved interviews with municipal representatives, multistory wood construction was perceived as a sustainable solution that enhances energy efficiency and permits quicker building with simple techniques, allowing the use of local renewable materials for the benefit of both buildings and local industry. [15]

While A. Takano And his colleagues revealed that wood for a residential building's structure lowered primary energy consumption throughout its life cycle by far more compared to other materials. For instance, the model house built using wood achieved a primary energy requirement of 310 MJ/m² per annum, as opposed to 370 MJ/m² for an equivalent one made of reinforced concrete.[16]

India
India is endowed with a tropical monsoon climate and huge seasonal fluctuations. The monsoon prevails in India, and the direction of the winds reverses completely from winter to summer. During winter, winds move from the northeast towards the southwest, providing dry and chilly conditions. During summer, the winds reverse and move from the southwest, bringing heat and humidity from the Indian Ocean.[17]

Earthen Plaster
According to F. Stazi And his team, earth plaster is a natural finishing material made mostly of fibers, sand, and clay used in the past to coat and cover earthen structures.[18]

In the study conducted by R. Paul, traditional earthen plasters applied to walls in India reduce indoor temperatures by 4 to 6°C in summer and maintain the indoor temperature 2 to 3°C above the outdoor temperature in winter. The plasters also have low thermal conductivity between 0.25 and 0.45 W/m·K, based on their composition (e.g., the content of plant fibers or ashes). This natural thermal performance limits the demand for air conditioning and heating systems, thereby reducing the energy consumption of traditional buildings by about 15 to 25% compared to modern uninsulated buildings. [19]

Lime
Lime is a centuries-old construction material that has made a resurgence due to its compatibility with architectural heritage preservation. It is produced from limestone through processes such as burning (calcination), slaking (water addition), ageing, and carbonation.[20]

Study [21] emphasizes the thermal behavior of lime plaster. Samples showed distinct mass losses at the various temperature steps, one surface-coat sample losing 12.7% between 30 and 600°C and 24.61% between 600 and 700°C; a thermal peak at ~123.8°C confirms gypsum dehydration, while its presence, along with hygroscopic zeolites, may help to retain moisture and improve thermal comfort.

The results indicate that lime plaster is a material that is both thermally stable and air-breathing and can effectively regulate internal temperature and humidity levels due to its mineral composition.

Mexico
Mexico's climate is highly diverse due to its extensive latitudinal range and the influence of subtropical high-pressure belts of the Northeast Pacific Ocean and Northern Atlantic. It is divided into two general climate regions by the Tropic of Cancer. The temperate climate of the north is generally arid or moderately moist, with mild to hot summers and cool to cold winters.[22]

The adobe
Adobe is a building material made predominantly of low-energy intensive and natural materials such as sand, soil, and cow dung, molded into blocks and dried in the sun.[23]

In the reference [24], the study evaluates the thermal performance of walls built with traditional Mexican materials with passive cooling methods like shading and insulation, the adobe was the best performing material regarding thermal performance. Adobe walls had a significantly lower thermal conductivity of 0.69 W/m·K and a higher thermal resistance of 0.88 m²·K/W. These characteristics allowed Adobe to successfully reduce heat transfer, with decreased indoor temperatures (less than 30°C), and improved thermal comfort and energy efficiency compared to the other materials like a solid block, partition block, and red brick.

The bamboo
Bamboo is one of the fast-growing, renewable, and low-cost materials available in most parts of the world, especially in tropical regions; some species attain a height of 36 m within six months.[25]

According to the article [26], Bamboo has 0.15–0.18 W/m·K thermal conductivity and is one of the most effective insulating materials, working perfectly in tropical climatic regions by lowering the indoor temperature up to 4°C, thereby cooling energy demands. Its light weight, natural insulation, and local availability create many reasons for bamboo to be an energy-efficient, eco-friendly choice, as shown in Cuilapam de Guerrero, Mexico.

Japan
Japan has a range of different climate conditions that changes from the sub-arctic climate to the subtropical climate, as there is a huge difference between both sides of Japan; the north has heavy snowfall, the east has warm and humid summers and cold winters, the west has very hot temperatures during summer months and moderate winter temperatures, and Okinawa is hot year around. [27]

Wood
The most used conventional building material in Japan, as the article suggests [28], is wood. Wood is valued for its ability to dampen natural forces, particularly earthquakes, due to its lightness and flexibility. Traditionally, Japanese buildings, such as homes and vertical buildings like pagodas and castles, were primarily constructed of wood over stone foundations.

The thatched roofs
Thatched roofs are vegetation material-based roofing systems. It is composed of grasses and palms, and used throughout the world as a traditional building practice. They are valued for their cultural significance and practical benefits, such as thermal insulation, temperature regulation, and heavy rainfall soundproofing. [29]

Another study about Thatched roofs of traditional Japanese houses, have an excellent thermal performance [30]. Because of the low thermal conductivity of thatch, in summer thatched attics are about 2°C cooler than outdoors during the day and 5-7°C cooler at night, whereas steel-roofed attics can reach more than 34°C, showing the better passive cooling of thatched roofs.

While M. Kokubo and his colleagues [31], the study aims to examine the thermal performance of Kyoto traditional Japanese house thatched roofs and the indoor microclimate. The researchers analyzed the evolution of the roof surface temperatures and indoor and outdoor humidity changes to establish the thermal efficiency of this type of construction. The main results are that the peak surface temperature of the thatch roof is around 60°C by daytime in summer, with the indoor temperature being more constant due to the insulating nature of the thatch.

Table 2 below provides a comparative summary of the thermal performance and energy efficiency contribution of traditional building materials in five nations,

Emerging Research in Materials for Environment, and Civil Infrastructure - GeoME 5.5   Materials Research Forum LLC
Materials Research Proceedings 58 (2026) 130-138   https://doi.org/10.21741/9781644903933-18

Table 2. *Relative analysis of energy efficiency and thermal performance of traditional building materials in five countries.*

Country	Material	Thermal Conductivity (W/m·K)	Indoor Temperature Reduction	Energy Demand Reduction	Other Key Performances
Morocco	Rammed Earth	-	-	Up to 32% reduction in heating demand	Reduced heat loss through walls and roofs
Morocco	Date Palm Fiber	0.243 (with 48% DPF)	-	Up to 25% reduction in cooling, 18% in heating	70% reduction in mortar thermal conductivity
Morocco	Stone (MPC)	U = 0.62 W/m²·K	<30°C in summer, >15°C in winter	-	Maintains humidity between 40% and 80%
Finland	Wood	0.12 – 0.14	-	Up to 38% heating energy savings	Reduces thermal bridges and improves airtightness
Finland	Wood (whole building)	-	-	310 MJ/m²/year vs. 370 MJ/m²/year (concrete)	69% reduction in embodied energy
India	Earthen Plaster	0.25 – 0.45	4 to 6°C reduction in summer	15% to 25% energy savings	Enhanced moisture control
India	Lime	-	-	-	Humidity regulation via calcination and water absorption
Mexico	Adobe	0.69	Indoor temperatures <30°C	Significant improvement in thermal comfort	Thermal resistance of 0.88 m²·K/W
Mexico	Bamboo	0.15 – 0.18	4°C reduction in indoor temp	Reduced use of air conditioning	Lightweight and good thermal insulator
Japan	Wood	0.12 – 0.14	-	-	High seismic resilience and rapid construction
Japan	Thatched Roof	Low (not specified)	Up to 7°C cooler indoors in summer	-	Significant thermal and sound insulation

Conclusions

This review demonstrates that traditional materials such as rammed earth, wood, bamboo, adobe, and thatch are highly thermally efficient in a wide variety of climates, thereby increasing comfort and lowering energy demand. Savings of up to 38% in Finland for heating with wood and of up to 25% in Morocco for cooling using date-palm fiber are examples. This study only utilized secondary data sources and did not consider hybrid systems that integrated traditional and modern techniques. Further research needs to be directed at the certification of new green materials and dynamic simulation to understand and optimize their performance in contemporary buildings better.

References

[1] E. Golden, Building from tradition: local materials and methods in contemporary architecture. London New York: Routledge, 2018.

[2] L. F. Guerrero Baca and F. J. Soria López, 'Traditional architecture and sustainable conservation', JCHMSD, vol. 8, no. 2, pp. 194–206, May 2018. https://doi.org/10.1108/JCHMSD-06-2017-0036

[3] D. G. Leo Samuel, K. Dharmasastha, S. M. Shiva Nagendra, and M. P. Maiya, 'Thermal com-fort in traditional buildings composed of local and modern construction materials', International Journal of Sustainable Built Environment, vol. 6, no. 2, pp. 463–475, Dec. 2017. https://doi.org/10.1016/j.ijsbe.2017.08.001

[4] 'Maroc_Etat_Climat_2023'.

[5] F. Ávila, E. Puertas, and R. Gallego, 'Characterization of the mechanical and physical properties of unstabilized rammed earth: A review', Construction and Building Materials, vol. 270, p. 121435, Feb. 2021. https://doi.org/10.1016/j.conbuildmat.2020.121435

[6] W. Cheikhi, 'Energy efficiency applied to vernacular architecture, case of the rammed earth buildings: A literature review', presented at the Vernacular Architecture: Support for Territo-rial Development, Feb. 2025, pp. 319–326. doi: 10.21741/9781644903391-37

[7] W. Cheikhi, K. Baba, S. M. Lamrani, A. Nounah, M. Khalfaoui, and L. Bahi, 'Study of indoor performances of a building using Rammed earth', MATEC Web Conf., vol. 149, p. 02089, 2018. https://doi.org/10.1051/matecconf/201814902089

[8] E. El-Kashif, S. Mehanny, and L. D., 'Effect of Environmental Conditions on Date Palm Fibers Composites', 2020, p. Chapter 11.

[9] S. Benaniba, Z. Driss, M. Djendel, E. Raouache, and R. Boubaaya, 'Thermo-mechanical characterization of a bio-composite mortar reinforced with date palm fiber', Journal of Engi-neered Fibers and Fabrics, vol. 15, p. 1558925020948234, Jan. 2020. https://doi.org/10.1177/1558925020948234

[10] M. Belhous, M. Boumhaout, S. Oukach, and H. Hamdi, 'Effect of a Material Based on Date Palm Fibers on the Thermal Behavior of a Residential Building in the Atlantic Climate of Mo-rocco', Sustainability, vol. 15, no. 7, Art. no. 7, Jan. 2023. https://doi.org/10.3390/su15076314

[11] M. Boumhaout, L. Boukhattem, H. Hamdi, B. Benhamou, and F. Ait Nouh, 'Thermomechanical characterization of a bio-composite building material: Mortar reinforced with date palm fibers mesh', Construction and Building Materials, vol. 135, pp. 241–250, Mar. 2017. https://doi.org/10.1016/j.conbuildmat.2016.12.217

[12] N. Laaroussi, N. Eraza, A. Hajji, and M. Mansour, 'Optimization of the Building Constructed of Confined Stone Walls According to the Moroccan Zoning', 2024, SSRN. https://doi.org/10.2139/ssrn.4823234

[13] H. Emre Ilgın and M. Karjalainen, 'Massive Wood Construction in Finland: Past, Present, and Future', in Wood Industry - Past, Present and Future Outlook, G. Du and X. Zhou, Eds., IntechOpen, 2023. https://doi.org/10.5772/intechopen.104979

[14] M. R. Cabral and P. Blanchet, 'A State of the Art of the Overall Energy Efficiency of Wood Buildings—An Overview and Future Possibilities', Materials, vol. 14, no. 8, Art. no. 8, Jan. 2021. https://doi.org/10.3390/ma14081848

[15] F. Franzini, R. Toivonen, and A. Toppinen, 'Why Not Wood? Benefits and Barriers of Wood as a Multistory Construction Material: Perceptions of Municipal Civil Servants from Finland', Buildings, vol. 8, no. 11, Art. no. 11, Nov. 2018. https://doi.org/10.3390/buildings8110159

[16] A. Takano, S. K. Pal, M. Kuittinen, and K. Alanne, 'Life cycle energy balance of residential buildings: A case study on hypothetical building models in Finland', Energy and Buildings, vol. 105, pp. 154–164, Oct. 2015. https://doi.org/10.1016/j.enbuild.2015.07.060

[17] D. N. K. Singh, 'Salient Features of Indian Climate'.

[18] F. Stazi, A. Nacci, F. Tittarelli, E. Pasqualini, and P. Munafò, 'An experimental study on earth plasters for earthen building protection: The effects of different admixtures and surface treatments', Journal of Cultural Heritage, vol. 17, pp. 27–41, Jan. 2016. https://doi.org/10.1016/j.culher.2015.07.009

[19] R. Paul, S. Girirajan, and S. Changali, 'Traditional Building Knowledge in Indian Lime and Earthen Plasters Conocimiento tradicional sobre revestimientos de tierra y cal en la India Conhecimentos tradicionais de construção nos rebocos de cal e terra da Índia', 2024.

[20] K. Elert, Rodriguez-Navarro ,Carlos, Pardo ,Eduardo Sebastian, Hansen ,Eric, and O. and Cazalla, 'Lime Mortars for the Conservation of Historic Buildings', Studies in Conservation, vol. 47, no. 1, pp. 62–75, Mar. 2002. https://doi.org/10.1179/sic.2002.47.1.62

[21] M. Singh and S. Vinodh Kumar, 'Mineralogical, Chemical, and Thermal Characterizations of Historic Lime Plasters of Thirteenth–Sixteenth-century Daulatabad Fort, India', Studies in Conservation, vol. 63, no. 8, pp. 482–496, Nov. 2018. https://doi.org/10.1080/00393630.2018.1457765

[22] 'Mexico.pdf'. Accessed: Mar. 18, 2025. [Online]. Available: https://files.cmcc.it/g20climaterisks/Mexico.pdf

[23] A. Shukla, G. N. Tiwari, and M. S. Sodha, 'Embodied energy analysis of adobe house', Renewable Energy, vol. 34, no. 3, pp. 755–761, Mar. 2009. https://doi.org/10.1016/j.renene.2008.04.002

[24] J. Uriarte-Flores, J. Xamán, Y. Chávez, I. Hernández-López, N. O. Moraga, and J. O. Aguilar, 'Thermal performance of walls with passive cooling techniques using traditional materials available in the Mexican market', Applied Thermal Engineering, vol. 149, pp. 1154–1169, Feb. 2019. https://doi.org/10.1016/j.applthermaleng.2018.12.045

[25] M. Fahim, M. Haris, W. Khan, and S. Zaman, 'Bamboo as a Construction Material: Prospects and Challenges', Adv. Sci. Technol. Res. J., vol. 16, no. 3, pp. 165–175, Jul. 2022. https://doi.org/10.12913/22998624/149737

[26] L. Á. R. Pérez and J. R. Ríos, 'Construcción de viviendas de interés social desde la perspectiva de los materiales sostenibles en Cuilapam de Guerrero, Oaxaca', Sapiens International Multidisciplinary Journal, vol. 1, no. 3, Art. no. 3, Dec. 2024. https://doi.org/10.71068/9px99j13

[27] 'Japan Meteorological Agency | General Information on Climate of Japan'. Accessed: Mar. 18, 2025. [Online]. Available: https://www.data.jma.go.jp/cpd/longfcst/en/tourist.html

[28] T. Porntavakool and N. Ongsavangchai, 'Japanese Characteristics in High-rise Buildings of Tokyo, Japan', Journal of Architectural/Planning Research and Studies (JARS), vol. 17, no. 2, Art. no. 2, Jun. 2020. https://doi.org/10.56261/jars.v17i2.239541

[29] C. Steger, 'A roof of one's own: choice and access in global thatch sustainability', World Development Sustainability, vol. 3, p. 100088, Dec. 2023. https://doi.org/10.1016/j.wds.2023.100088

[30] H. Yoshino, K. Hasegawa, and S. Matsumoto, 'Passive cooling effect of traditional Japanese building's features', Management of Env Quality, vol. 18, no. 5, pp. 578–590, Aug. 2007. https://doi.org/10.1108/14777830710778337

[31] M. Kokubo et al., 'Relationship between the Microbiome and Indoor Temperature/Humidity in a Traditional Japanese House with a Thatched Roof in Kyoto, Japan', Diversity, vol. 13, no. 10, p. 475, Sep. 2021. https://doi.org/10.3390/d13100475

Emerging Research in Materials for Environment, and Civil Infrastructure - GeoME 5.5   Materials Research Forum LLC
Materials Research Proceedings 58 (2026) 139-146   https://doi.org/10.21741/9781644903933-19

A comprehensive life cycle assessment of sustainable reinforcement solutions for expansive soils using natural materials

Ahlam EL MAJID[1,*], Khadija BABA[1], Latifa OUADIF[2], Yasmina ED-DARIY[3]

[1]Civil and Environmental Engineering Laboratory (LGCE), Mohammadia Engineering School, Mohammed V University in Rabat, Morocco

[2]Laboratory of Applied Geophysics, Geotechnics, Engineering Geology, and Environment (L3GIE), Mohammadia Engineering School, Mohammed V University in Rabat, Morocco

[3]National School of Architecture, Fez, Morocco

*ahlam.elmajid@research.emi.ac.ma

https://orcid.org/0000-0001-8242-1623

Keywords: Expansive Soils, Geotechnical Issues, Soil Stabilization, Natural Fibers, Bio-Based Reinforcement, Sustainable Alternatives

Abstract. Expansive soils cause serious geotechnical issues through their moisture-sensitive volume changes, the results of which are structural integrity loss. Even though traditional stabilization methods using lime or cement are effective, their significant carbon footprint as well as non-renewable nature raise environmental issues. This study conducts a comprehensive life cycle assessment to forecast the long-term viability of bio-based expansive soil stabilizers, pitting natural fibres (*Alfa, jute, sisal*), biopolymers, and agro-industrial by-products against conventional methods. The Life Cycle Assessment (LCA) approach examines key environmental factors, embodied carbon, energy use, resource use, and long-term durability. The results show that using natural stuff to strengthen soil is good for the earth. The implementation of these materials can lower carbon emissions by 40–60% compared to cement. These materials maintain control over soil swelling and aid in improving soil strength. Agricultural by-products are cheap, can be renewed, and don't need much work to use. A chart will compare how well these materials do over their lives. This chart can aid decision-makers in making sustainable choices. Also, using local plant-based resources helps local economies and supports green building work. However, large-scale implementation remains challenging due to limited availability, inconsistent fiber quality, and the lack of standardized guidelines. Research is finding optimal combinations of fibers and new ways to prepare them to be both strong and good for the environment. This reduces ecological footprints without sacrificing performance. to adopt low-impact stabilization. Future studies must focus on large-scale implementation and long-term field performance for gaining universal industry application.

1. Introduction

Soils expansive in nature present significant problems, particularly in arid areas [1,2]. Conventional stabilization techniques (like lime or cement) successfully address these issues but with high energy input and CO_2 emissions, further accelerating environmental degradation [3]. With growing anxiety over the environmental footprint of construction work, there has been rising demand for environmentally friendly, low-carbon alternatives. Natural fibers, biopolymers, and crop residues have become increasingly recognized due to their renewable nature and local availability. Plant based stabilizers have clear environmental and social pluses, giving us a plan for greener ways to work with soil. To make things better, the research consider the whole life cycle. plant-based stabilizers to see how good they are when they pertain to the environment. This study takes a look at carbon use and energy in addition to how much we use natural resources.

Emerging Research in Materials for Environment, and Civil Infrastructure - GeoME 5.5   Materials Research Forum LLC
Materials Research Proceedings 58 (2026) 139-146   https://doi.org/10.21741/9781644903933-19

2. Materials and Methods.

This study uses a detailed experimental way of looking at how plant-based soil stabilizers act in soils that expand. The method has lab tests and a view of the whole life cycle. It gives information on both technical and sustainability topics.

2.1 Selection and Preparation of Materials

Three types of plant based stabilizers were chosen to compare:

2.1.1 Natural fibers

Natural fibers have different things about them that change how they help stabilize soil. The three fibers weigh about the same, from 1.3 to 1.4 grams per cubic centimeter. Alfa fibers are thin, from 5 to 22 micrometers. Jute and Sisal fibers are a bit thicker, measuring 15 to 35 micrometers and 10 to 20 micrometers, respectively. How much they stretch differs: Alfa stretches more (1.4% to 5%), while Jute (1.5% to 1.8%) and Sisal (2% to 2.5%) stretch less. Alfa fibers are also the strongest, from 173.4 to 1327 MPa, then Jute (400 to 800 MPa) and Sisal (511 to 635 MPa). How stiff they are, which is measured by Young's modulus, ranges from 18 to 58 GPa for Alfa fibers, from 10 to 30 GPa for Jute, and from 9.4 to 22 GPa for Sisal. How much cellulose is in them, which is one of the main things that makes them strong, is highest in Jute and Sisal (67–71.5% and 67–78%, respectively) and a bit less in Alfa (38.8–47.6%). Lignin content is more higher in Alfa (14.9–24%) compared to Jute (12–13%) and Sisal (8–11%). The microfibrillar angle, which affects elasticity, is around 8° for Jute and 11° for Sisal, while Alfa's value is not commonly reported. Wax content remains relatively low across all fibers, ranging from 0.5% in Jute to 2% in Sisal. Hemi-cellulose content is highest in Alfa (22.1–38.5%), followed by Jute (13.6–20.4%) and Sisal (10–14.2%).

2.1.2 Agro-industrial byproducts (Fly ash).

All stabilizing agents were sourced locally where possible to reflect realistic supply chains and minimize transport emissions. Fibers were cut into standardized lengths (10–20 mm), and dried to improve bonding with soil particles, the proprieties of Fly-Ash are presented in Table 1.

Table 1. the proprieties of Fly-Ash

Property	Value
Color	Gray to dark gray
Specific Gravity	2.1 – 2.6
Fineness (Retained on 45 μm) (%)	< 34
Particle Size (mm)	< 0.075
pH	6 – 12
Loss on Ignition (LOI) (%)	< 6 (Class F), < 10 (Class C)
SiO_2 (%)	35 – 60
Al_2O_3 (%)	10 – 30
Fe_2O_3 (Iron Oxide)(%)	4 – 20
CaO (%)	< 10 (Class F), > 20 (Class C)
Moisture Content(%)	< 1.5
Pozzolanic Activity Index(%)	≥ 75
Bulk Density(kg/m³)	800 – 1100

2.2 Soil Characterization and Stabilization Procedure

A highly plastic expansive clay was selected as the base soil. The geotechnical characteristics of the samples are summarized in Table 2 below:

Emerging Research in Materials for Environment, and Civil Infrastructure - GeoME 5.5   Materials Research Forum LLC
Materials Research Proceedings 58 (2026) 139-146   https://doi.org/10.21741/9781644903933-19

Table 2. the proprieties of clay soil [4]

Parameter	Details	Value
particle size analysis MS 13.1.008	%< 0,08 mm	93.7
	%< 2mm	98.7
	%<20mm	100
Water content w(%)	-	16.6
Atterberg limits	Liquid limit LL(%)	62
	plasticity index PI (%)	38
Voluminal mass γ (kg/m3)	-	1830
Dry voluminal mass γd(kg/m3)	-	1740
Classification	-	A3
Free swelling index (%)	-	90
Optimum Moisture Content, %	-	30.7
CBR %	Unsoaked	8.27
	Soaked	3.25
Unconfined Compressive Strength kN/m2	-	70.5
Cohession, kN/m2 (uu Test)	-	35.2

Stabilized soil specimens were prepared by mixing the expansive clay with varying contents (0.5% to 2% by weight) of each stabilizer. The mixtures were cured under controlled conditions for 7, 14, and 28 days to evaluate short- and medium-term effects.

2.3 Life Cycle Evaluation
The effects on the environment of each soil stabilization method were assessed using a complete environmental impact evaluation from extraction of raw materials to disposal at the end of life, using the internationally recognized *ISO 14040* and *ISO 14044* standards framework for Life Cycle Assessment. The analysis focused on treating one cubic meter *(1 m³)* of expansive soil as the functional unit. The study covered all stages from extraction of raw materials, processing, and transportation to application and final disposal or end-of-life management.

Key impact categories evaluated include:
- Global Warming Potential (*GWP*), which expresses the amount of CO_2 equivalent emitted and is measured in kilograms (kg CO_2eq)
- How much total energy is needed (*CED*), in megajoules (*MJ*)
- How much we use up resources that can't be regrown.
- Harm to people and the earth.

The LCA was realised using *SimaPro v9.4* with the *Ecoinvent 3.8* database.

2.4 Comparative Analysis
Both sets of data, dealing with the environment and building, were made standard and put on a performance matrix that looks at many things. This allows for comparing how much stronger something is to how it affects the environment. A test was done to see how much the amount of stabilizer and ways of treating it affect how things perform for the environment as a whole.

3. Results

3.1 Swelling Potential
This study examined what occurs when plant fibers are included in two soil types. How much fiber you add and how much the soil swells depend on each other. Adding more fiber from 0.5% to 2% had better results.
- Clay soils showed 12.25% less swelling (average 3.14%)

- Marl soils exhibited 10.23% reduction (average 3.52%)

The amount of time the soil was wet didn't change how much it swelled. With the best amount of fiber (2% fiber and 25 mm fibers), the swelling stayed the same over time in all soils. This tells us that the treatment stabilizes things for a long time. Since what happens depends on the soil, you need to find answers that fit each site for geotechnical design. While lab results are good, real-world tests under different conditions would be good. Figure 1 shows these numbers, making clear how adding fiber holds back soil swelling in both clay and marl.

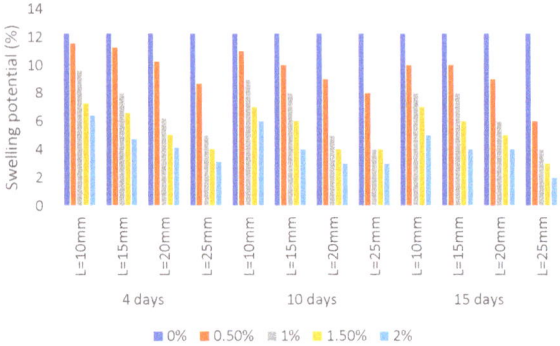

(a)

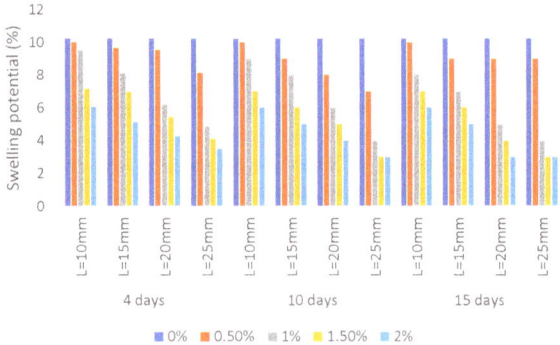

(b)

Fig. 1. Effect of Swelling potential (%) in expansive soil (a) Clay, (b) Marl

3.2 Swelling Pressure
Our test showed that polypropylene bound fibers can lower swelling pressure in two soil types.
(a) Clay Soil Results:
- Unreinforced: 72 kPa swelling pressure
- Using 2% brings the pressure down to 44.5 kPa, dropping it by over 38%. The length is important because 25 mm of fibers worked better than just ten millimeters (44.5 kPa).

(b) Marl Soil Results:

Emerging Research in Materials for Environment, and Civil Infrastructure - GeoME 5.5   Materials Research Forum LLC
Materials Research Proceedings 58 (2026) 139-146   https://doi.org/10.21741/9781644903933-19

- Unreinforced: 121 kPa swelling pressure
- When you add 2%, the pressure goes to only 89 kPa.

The data is clear:

1. Adding more fiber (from 0.5% to 2%) lowers pressure more.
2. Longer fibers (25 mm) work better.
3. Clay soils react to stabilization better than marl soils. Lab tests say that adding polypropylene fibers works. The swelling pressure drops the most in clay soils that tend to swell. When you raise the amount and length of fiber, the results get better. This means you can trust this method to do well if the right fiber length is used.

Using fiber reinforcement is a solid alternative to traditional stabilization techniques. Clay soils gain more from this than marl soils, as seen in Figure 2.

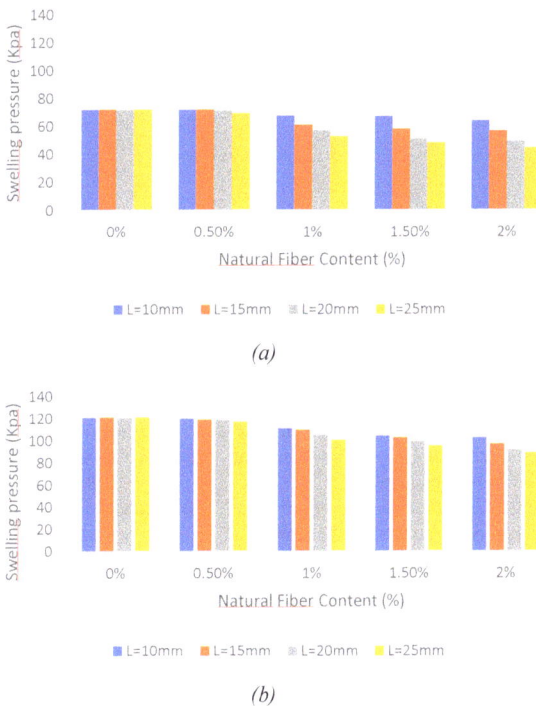

(a)

(b)

Fig. 2. Effect of Swelling pressure in Different Soil Types: (a) Clay Sample, (b) Marl Sample

3.3 Life Cycle Assessment (LCA)

When stabilization methods are compared, natural fibers are best. They do better than fly ash solutions on every environmental measure. Natural fiber stabilization has a global warming potential (GWP) of just 48 kg of CO_2 per cubic meter. Compared to the fly ash method, that is a 60% drop. The fly ash method rests at 120 kg. Similarly, the cumulative energy requirement (CED) also decreases from 950 MJ/m³ for Fly-ash to 380 MJ/m³ for natural fibers, indicating an energy-efficient process. The abiotic resource depletion value of natural fibers (0.007) is also significantly lower than that of fly-ash (0.018), indicating lower reliance on non renewable resources. In terms of water consumption, natural fibers consume 125 liters per cubic meter, while fly-ash consumes 210 liters, a 40% decrease. Lastly, the human and ecosystem toxicity index falls from 3.2 to 1.1, favoring the environmental and health benefits of using bio-based products. Recycled polypropylene fibers also have an environmental improvement over fly-ash, though to a lesser extent than with natural fibers.

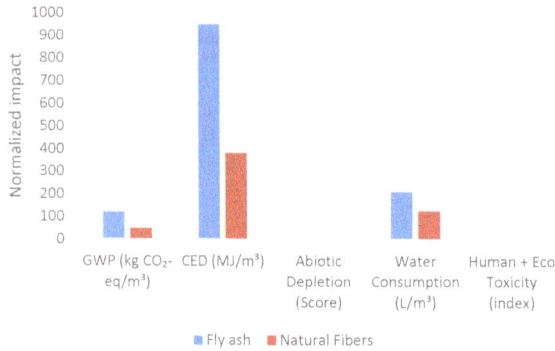

Fig. 3. Normalized Environmental Impacts of Soil Stabilization Methods per 1 m^3 of Treated Expansive Soil

4. Discussion

Soil expansion maintenance remains one of the major challenges in geotechnical engineering, as subsurface stability is typically the deciding factor in the long-term performance and longevity of structures. To provide reliable results, one has to be aware of the long-term behavior of fiber-treated soils under varying moisture and temperature conditions.

Further work would then be focused on the long-term response of stabilized soils, with most significant factors such as soil mineralogy and the characteristics of different fibers. These are significant in the development of consistent and pragmatic guidelines for the design and construction of fiber-reinforced soils.

fiber reinforcement is an effective solution and eco-friendly way to reduce swelling in expansive soils. Faisal Noaman et al. and Soltani et al. [5,6] have, for example, determined in their research that the incorporation of non-expansive fibers enhances dimensional stability by limiting water-induced expansion. El Majid et al. [7–10] also demonstrated that longer fibers employed in marl soils lead to greater swelling reductions. Shukla et. [11] also provided helpful guidance on the application of fiber-reinforced soils in real geotechnical applications with emphasis on careful consideration of parameters affecting behavior to achieve maximum performance.

The researches coupled confirm that fiber reinforcement is a viable and efficient method of preventing swelling in expansive soils, which allows the construction of more resistant and long-lasting foundations for civil engineering purposes.

Conclusion

The conclusions of this study confirm that natural fiber reinforcement of expansive soils drastically reduces swelling capacity and improves overall stability. Higher content of fibers always leads to reduced swelling pressures, and therefore the determination of the best dosages for each type of soil becomes important. Furthermore, longer fiber forms tend to offer higher benefits, and therefore proper fiber length selection for improving reinforcement efficiency becomes imperative.

Emerging Research in Materials for Environment, and Civil Infrastructure - GeoME 5.5   Materials Research Forum LLC
Materials Research Proceedings 58 (2026) 139-146   https://doi.org/10.21741/9781644903933-19

This environmentally friendly technique has significant advantages for geotechnical engineering, where long-term structural integrity is critical. With the use of natural fiber reinforcement, engineers are able to design more stable and sustainable stabilization systems. More extensive future studies should investigate the long-term behavior of fiber-reinforced soils under true field conditions and explore other types of natural and hybrid fibers. Such studies will

References

[1] Chen, F. (2012). Foundations on expansive soils.
https://books.google.com/books?hl=fr&lr=&id=C30h1HpQvDUC&oi=fnd&pg=PP1&dq=Chen,
+F.H.+(1988).+Foundations+on+Expansive+Soils.+Elsevier.&ots=0WSK9CxboC&sig=sH0KN
4YWIC0ZyNXsIprDRFsik1I

[2] Nelson, J., & Miller, D. (1997). Expansive soils: problems and practice in foundation and pavement engineering
https://books.google.com/books?hl=fr&lr=&id=ISvJ33HHxDEC&oi=fnd&pg=PR17&dq=Nelso
n,+J.D.,+%26+Miller,+D.J.+(1992).+Expansive+Soils:+Problems+and+Practice+in+Foundation
+and+Pavement+Engineering.+Wiley.&ots=ogqsMEZWr_&sig=xvNXROnDnrJlqKsBdVDenq
Cm-pQ

[3] Senadheera, S., Withana, P., Lim, J., … S. Y.-G., & 2024, undefined. (n.d.). Carbon negative biochar systems contribute to sustainable urban green infrastructure: a critical review. Pubs.Rsc.Org. Retrieved June 16, 2025, from
https://pubs.rsc.org/en/content/articlehtml/2024/gc/d4gc03071k

[4] El Majid, A., Baba, K., & Razzouk, Y. (2023a). Assessing the impact of plant fibers on swelling parameters of two varieties of expansive soil. Case Studies in Chemical and Environmental Engineering, 8, 100408. https://doi.org/10.1016/J.CSCEE.2023.100408

[5] Faisal Noaman, M., Khan, M. A., & Ali, K. (2022). Effect of artificial and natural fibers on behavior of soil. Materials Today: Proceedings, 64, 481–487.
https://doi.org/10.1016/J.MATPR.2022.04.954

[6] Soltani, A., Taheri, A., Khatibi, M., & Estabragh, A. R. (2017). Swelling Potential of a Stabilized Expansive Soil: A Comparative Experimental Study. Geotechnical and Geological Engineering, 35(4), 1717–1744. https://doi.org/10.1007/S10706-017-0204-1/METRICS

[7] El Majid, A., Baba, K., & Razzouk, Y. (2023b). Mitigating Soil Swelling: Exploring the Efficacy of Polypropylene Fiber Reinforcement in Controlling Expansion of Expansive Soils. Springer Proceedings in Earth and Environmental Sciences, Part F1971, 183–192.
https://doi.org/10.1007/978-3-031-49345-4_19

[8] EL MAJID, A., CHERRADI, C., BABA, K., & RAZZOUK, Y. (2023). Laboratory investigations on the behavior of CBR in two expanding soils reinforced with plant fibers of varying lengths and content. Materials Today: Proceedings.
https://doi.org/10.1016/J.MATPR.2023.06.395

[9] Majid, E. A., Kh, B., Ya, R., El Majid, A., Baba, K., & Razzouk, Y. (2024). Сравнительный анализ снижения набухания мергельно-глинистых грунтов: натуральные растительные волокна (Alfa, джут, сизаль) против полипропиленовой фибры с известково-пуццолановым цементом при прокторном уплотнении. Bulletin of the Tomsk Polytechnic University Geo Assets Engineering, 335(4), 52–63.
https://doi.org/10.18799/24131830/2024/4/4410

[10] El Majid, A., Baba, K., & Razzouk, Y. (2025). Expansive Soils Stabilized with Recycled Polypropylene Fibers: An Assessment Based on Laboratory and Life Cycle Data for Mechanical and Environmental Performance. *Engineering Proceedings, 112*(1), 28. https://doi.org/10.3390/engproc2025112028

[11] Shukla, S. K. (2017). Fundamentals of Fibre-Reinforced Soil Engineering. https://doi.org/10.1007/978-981-10-3063-5

Emerging Research in Materials for Environment, and Civil Infrastructure - GeoME 5.5   Materials Research Forum LLC
Materials Research Proceedings 58 (2026) 147-155   https://doi.org/10.21741/9781644903933-20

Bonding mechanism of hot-pressed green composites using FTIR spectroscopy

Ejazulhaq Rahimi[1*], Ayane Yui[1], Yuta Yamachi[1], Yuma Kawasaki[1]

[1]Graduate School of Science and Engineering, Ritsumeikan University, Kyoto, Japan

ejaz.rahimi@outlook.com

Keywords: Recycled Concrete Powder, Hemicellulose, Cellulose, Lignin, FTIR

Abstract. CO_2 emission and solid waste generation from cement and concrete constitute major environmental challenges in construction industry. A significant portion of the emission stems from the production of clinker, the primary binder in conventional concrete. Consequently, considerable research efforts have been directed toward the development of environmentally friendly alternative binders. Among these, hemicellulose combined with recycled concrete powder (RCP) processed by hot-pressing has shown great flexural strength. However, the bonding mechanism within the composite is not studied. This study aims to investigate the bonding mechanism in the composite using FTIR. The composite specimens were fabricated at 24 °C and 60 °C of pressing temperatures. Results showed no evidence of chemical reactions. Instead, the hardening is primarily attributed to surface interactions through hydrogen bonding.

Introduction

Construction industry is one of the major contributors to global solid waste generation and CO_2 emission. Cement calcination accounts for 7-8% of CO_2 emission, making it a significant driver of the greenhouse effect and climate change [1]. Additionally, construction and demolition activities produce a huge quantity of solid waste, presenting another major environmental challenge. As of 2012, the total construction and demolition waste generated across 40 countries exceeded 3 billion tonnes annually [2]. Recycled concrete aggregate (RCA) has been employed in the production of new concrete, however, its mechanical performance remains inferior to that of natural aggregate, exhibiting 10-20% lower strength, which continues to decline with each successive recycling cycles [3-5]. Furthermore, the current recycling methods requires new cement to produce concrete, which indicates that the complete of cycle of recycling is still elusive.

This study is from a series of studies exploring the development of a composite materials from RCP and wood-based biomass components. Previous studies showed that composites made of RCP and hemicellulose achieved superior flexural strength compared to the other biomass-based composites [6]. However, the bonding mechanism underlying the strength improvement remains unclear.

This study is to unveil the nature of bonding between RCP and the biomass components – specifically cellulose, lignin, and hemicellulose - using FTIR. It is employed to assess the presence of chemical interactions and the formation of new phases. It worth mentioning that this research is the first comprehensive spectroscopic investigation of RCP-biomass composites fabricated by hot-pressing. The findings are expected to contribute significantly to the achievement of non-cementitious, eco-friendly construction materials technologies.

Experimental Program

Materials and Sample Preparation. In this study, the raw materials are RCP, cellulose, lignin, hemicellulose, oyster shell and chitosan. The specimens were prepared using the preparation method applied in our previous work [6]. After curing the specimens were finely ground for the chemical examination.

Emerging Research in Materials for Environment, and Civil Infrastructure - GeoME 5.5   Materials Research Forum LLC
Materials Research Proceedings 58 (2026) 147-155   https://doi.org/10.21741/9781644903933-20

Experimental Procedure. FTIR spectroscopy was conducted using a PerkinElmer Spectrum 3 spectrometer equipped with a Spotligh 400i (MIR/NIR) system (PerkinElmer, Japan). The test was carried out in the range of 4000-400 cm-1 at a resolution of 4 cm-1. For comparative analysis and consistency of the results, the spectra were normalized to the absorption peak at 875 cm-1, attributed to the out-of-plane vibration of the C-O bond, which is one of the characteristic peaks of the RCP.

Results and Discussion
FTIR analysis was conducted to unveil the bonding mechanism of the composite incorporating RCP and hemicellulose as well as cellulose and lignin synthesized at 24 °C and 60 °C.

Fig. 1 illustrates the FTIR spectra of RCP, cellulose, and their composites processed at both 24 °C and 60 °C. an absorption band centered at around 3437 cm-1 is evident in the composite spectrum, corresponding to O H stretching vibrations. These vibrations are attributed to groups of hydroxyl in cellulose and water associated with portlandite (Ca(OH)2, which is generated during the hydration of C3S and C2S, as well as from hydrogen-bind O-H groups adsorbed on the surfaces of cementitious particles [7]. A peak at 2917 cm 1is assigned to C-H stretching vibration in cellulose [8].

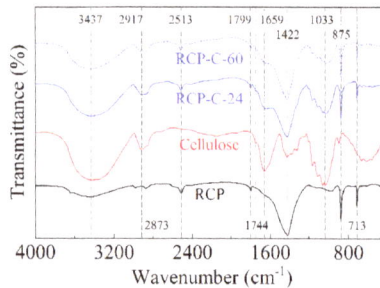

Fig. 1. FTIR spectra of RCP, cellulose, and their composite prepared at 24 °C and 60 °C

The peak at 2513 cm^{-1} is associated with the carbonate ion (CO_3^{2-}) [9, 10], while the distinct peak at 1799 cm-1 corresponds to the combination band of CO_3^{2-} ions in calcium carbonate ($CaCO_3$) [10]. The band observed at 1659 cm^{-1} is attributed to the H-O-H bending vibration, characteristic of water in cellulose [11]. The peak at 1422 cm^{-1} can be attributed to both the C-H bending vibration in cellulose and the symmetric stretching of the carboxylate (COO-) groups in $CaCO_3$ derived from RCP [12, 13]. A strong peak at 1033 cm^{-1} is representative of C-O stretching vibrations within the cellulose polymer backbone and Si-O stretching modes associated with silicate structures in RCP. The peak observed at 875 cm^{-1} corresponds to the out-of-plane bending vibration of CO_3^{2-} ions in $CaCO_3$ [14], whereas the peak at 713 cm^{-1} is linked to the in-plane bending vibration of CO_3^{2-} ions [14].

The summary of the FTIR band from the composite made of RCP and cellulose is presented in Table 1.

Emerging Research in Materials for Environment, and Civil Infrastructure - GeoME 5.5   Materials Research Forum LLC
Materials Research Proceedings 58 (2026) 147-155   https://doi.org/10.21741/9781644903933-20

Table 1. FTIR spectra of the composite made of RCP and cellulose

Wavenumber [cm-1]	Functional Bond	Assigned	Reference
3437	O-H stretching	Hydroxyl groups in cellulose and water in portlandite	[7]
2917	C-H stretching (CH2)	Cellulose	[8]
2513	2-	Carbonate ion in CaCO3	[9], [10]
1799	2-	Carbonate ion in CaCO3	[10]
1659	H-O-H bending	Water molecules associated with cellulose	[11]

Fig. 2 displays the FTIR spectra of RCP, lignin, and their composites fabricated at pressing temperatures of 24 °C and 60 °C. The spectra of the composite largely represent a superimposition of the characteristic peaks from both RCP and lignin, with no discernible emergence of new absorption bands or significant shifts in existing peaks positions. This spectral behavior suggests the absence of substantial chemical interactions or the formation of new chemical bonds between lignin and RCP under the given processing conditions.

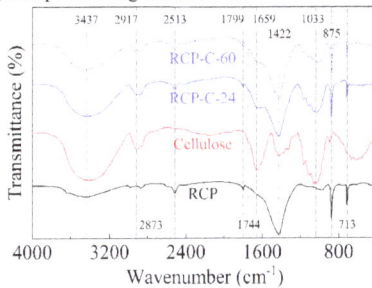

Fig. 2. FTIR spectra of the composite made of RCP and lignin prepared at 24 °C and 60 °C

The peak observed centering at approximately 3450 cm^{-1} is attributed to the O-H stretching vibrations associated with hydroxyl groups in lignin and the moisture content within RCP [7, 15]. The peak at 2936 cm^{-1} corresponds to the asymmetric C-H stretching of $-CH_2$ and $-CH_3$ groups in lignin [16], while the band at 2841 cm^{-1} is ascribed to symmetric C–H stretching vibrations [17]. The absorption bands at 2513 cm^{-1} and 1799 cm-1 are indicative of CO_3^{2-} in $CaCO_3$ [9], [10].

The band at 1423 cm^{-1} is associated with the symmetric stretching vibration of CO_3^{2-} groups is $CaCO_3$, coupled with C-H bending vibrations in lignin [14], [19]. The peak at 1266 cm^{-1} corresponds to C-O stretching within the guaiacyl ring of lignin [20], while the band at 1213 cm^{-1} is assigned to C-O-C stretching vibrations typical of lignin's ether linkages [21]. The absorption at 1128 cm^{-1} is linked to both C O and C-H stretching vibrations within the guaiacyl unit [22]. The peak at 1039 cm^{-1} results from the stretching vibrations of primary aliphatic hydroxyl groups in lignin and overlapping Si-O stretching in the silicate matrix of RCP [10, 14]. Finally, the absorption peaks at 875 cm^{-1} and 713 cm^{-1} correspond to the out-of-pane bending mode of CO_3^{2-}, respectively, signifying the presence of $CaCO_3$ in the composite structure [14].

Emerging Research in Materials for Environment, and Civil Infrastructure - GeoME 5.5   Materials Research Forum LLC
Materials Research Proceedings 58 (2026) 147-155   https://doi.org/10.21741/9781644903933-20

The summary of the FTIR bands from the composite made of RCP and lignin is presented in Table 2.

Table 2. FTIR spectra of the composite made of RCP and lignin

Wavenumber [cm⁻¹]	Functional Bond	Assigned to	Reference
3450	O–H stretching	Hydroxyl groups in lignin and water in RCP	[7], [15]
2936	C–H asymmetric stretching	–CH_2/–CH_3 groups in lignin	[16]
2841	C–H symmetric stretching	Lignin	[17]
2513	CO_3^{2-} vibration	Carbonate ion in $CaCO_3$	[9], [10]
1799	Combined mode (CO_3^{2-})	Carbonate ion in $CaCO_3$	[10]
1596	Aromatic skeletal vibration	Lignin	[18]
1423	Symmetric CO_3^{2-} stretch / C–H bending	Carbonate ion in $CaCO_3$ and lignin	[14], [19]
1266	C–O stretching (guaiacyl ring)	Lignin	[20]
1213	C–O–C stretching	Lignin	[21]
1128	C–H and C–O stretching	Guaiacyl unit in lignin	[22]
1039	Aliphatic –OH stretching / Si–O stretching	Primary alcohol groups in lignin and silicates in RCP	[10], [14]
875	Out-of-plane bending (CO_3^{2-})	Carbonate ion in $CaCO_3$	[14]
713	In-plane bending (CO_3^{2-})	Carbonate ion in $CaCO_3$	[14]

Fig. 3 represents the FTIR spectra of RCP, hemicellulose, and their corresponding composites fabricated at pressing temperatures of 24 °C and 60 °C. the spectra of the RCP-H-24 and RCP-H-60 composites generally represent a superimposition of the individual spectra of RCP and hemicellulose, indicating an absence of significant chemical interaction or the formation of new functional groups. However, the characteristic absorption bands of hemicellulose appear considerably attenuated in the composite spectra when compared to those in the cellulose and lignin-based composites.

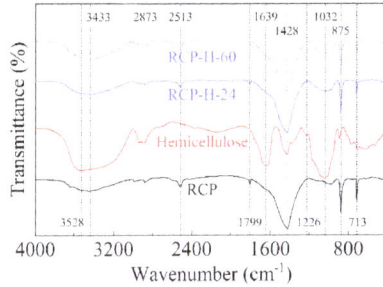

Fig. 3. FTIR spectra of the composite made of RCP and hemicellulose prepared at 24 °C and 60 °C

In particular, the intensity of the broad absorption band associated with O-H stretching vibrations in hemicellulose is markedly diminished in the composite spectra. A similar reduction is observed for the band at 1639 cm^{-1}, which is typically assigned to the bending vibrations of absorbed water of C=O stretching in hemicellulose. Furthermore, the broad band centered around 1032 cm^{-1}, attributed to C-O and C-C stretching vibrations within the hemicellulose backbone, also shows a significant reduction in intensity. However, the specimens using cellulose and lignin display less attenuation in their characteristic peak intensities.

Furthermore, the spectra of RCP-H-24 and RCP-H-60 exhibit similar profiles, with the high peak intensities appearing at the same wavenumbers, suggesting that the pressing temperature does not induce any significant chemical interactions between RCP and hemicellulose.

The peak intensity at 3528 cm^{-1} associates to O-H stretching vibrations corresponding to hydroxyl groups [23], while the band at 3433 cm^{-1} corresponds to O-H stretching in hemicellulose and moisture existed in RCP [24, 25]. The peak at 2873 cm^{-1} is associated to C-H stretching vibrations [26].

The peak observed at 2513 cm^{-1} is assigned to CO_3^{2-}, and the band at 1799 cm^{-1} represents the combined vibrational mode of CO_3^{2-} in $CaCO_3$ [9, 10]. The absorption at 1639 cm^{-1} is indicative of H-O-H bending vibrations, consistent with the high-water affinity of holocellulose [27].

The band at 1428 cm^{-1} is attributed to CH2 bending vibrations present in both hemicellulose and cellulose structures [28]. The absorption at 1226 cm^{-1} is linked to C-C and C-O-C stretching vibrations typically found in lignin but may appear due to trace amounts in hemicellulose [29]. The prominent peak at 1032 cm^{-1} corresponds to C-O stretching vibrations of cellulose and primary alcohol groups, along with in-plane C H deformation in the guaiacyl unit [30]. Finally, the peaks at 875 cm^{-1} and 713 cm^{-1} are characteristic of the out-of-plane and in-plane bending modes of CO_3^{2-} in $CaCO_3$, respectively [14].

The summary of the FTIR bands from the composite made of RCP and hemicellulose is presented in Table 3.

Table 3. FTIR spectra of the composite made of RCP and hemicellulose

Wavenumber [cm^{-1}]	Functional bond	Assigned to	Reference
3528	O–H stretching	Hydroxyl groups	[23]
3433	O–H stretching	Hemicellulose and moisture in RCP	[24], [25]
2873	C–H stretching	Composite material	[26]
2513	CO_3^{2-} vibration	Carbonate ion	[9], [10]
1799	Combined CO_3^{2-} mode	$CaCO_3$	[10]
1639	H_2O bending	Absorbed water in holocellulose	[27]
1428	CH_2 bending	Cellulose and hemicellulose	[28]
1226	C–C and C–O–C stretching	Lignin	[29]
1032	C–O stretching / C–H in-plane deformation	Cellulose, primary alcohols, and guaiacyl unit	[30]
875	Out-of-plane CO_3^{2-} bending	$CaCO_3$	[14]
713	In-plane CO_3^{2-} bending	$CaCO_3$	[14]

Conclusion

This study presented a chemical characterization of green composites synthesized from RCP and biomass derived components-namely cellulose, lignin, and hemicellulose-using FTIR spectroscopy. The primary aim was to elucidate the interaction mechanisms between RCP and biopolymers during composite formation. FTIR analysis revealed that the dominant interaction between RCP and the biomass binders was governed by surface-level hydrogen bonding rather than chemical reaction or new phase formation. Among cellulose, lignin, and hemicellulose, the later showed the strongest bond RCP, seen by a greater shifts and reduction of intensity in its FTIR peaks. Cellulose also exhibited significant interaction, but to a lesser extend, while lignin showed the weakest bond. Overall, the results provide great insight into the surface bonding behavior between RCP and biomass binders and contribute to the growing body of knowledge necessary for the development of fully recycled, non-cementitious green construction materials.

Acknowledgements

The financial support of RARA Office at Ritsumeikan University is sincerely acknowledged, as it helped the successful completion of this research.

References

[1] S. Barbhuiya, F. Kanavaris, B. B. Das, and M. Idrees, "Decarbonising cement and concrete production: Strategies, challenges and pathways for sustainable development," *Journal of Building Engineering*, vol. 86, p. 108861, Jun. 2024. https://doi.org/10.1016/j.jobe.2024.108861

[2] A. Akhtar and A. K. Sarmah, "Construction and demolition waste generation and properties of recycled

aggregate concrete: A global perspective," *J Clean Prod*, vol. 186, pp. 262–281, Jun. 2018. https://doi.org/10.1016/J.JCLEPRO.2018.03.085

[3] P. Zhu, X. Chen, H. Liu, Z. Wang, C. Chen, and H. Li, "Recycling of waste recycled aggregate concrete in

freeze-thaw environment and emergy analysis of concrete recycling system," *Journal of Building Engineering*, vol. 96, p. 110377, Nov. 2024. https://doi.org/10.1016/j.jobe.2024.110377

[4] W. K. Dong, W. G. Li, and Z. Tao, "A comprehensive review on performance of cementitious and geopolymeric concretes with recycled waste glass as powder, sand or cullet," *Resour Conserv Recycl*, vol. 172, 2021. https://doi.org/10.1016/j.resconrec.2021.105664

[5] J. Kim and A. Ubysz, "Thermal activation of multi-recycled concrete powder as supplementary cementitious

material for repeated and waste-free recycling," *Journal of Building Engineering*, vol. 98, p. 111169, Dec. 2024. https://doi.org/10.1016/j.jobe.2024.111169

[6] E. Rahimi, Y. Kawasaki, A. Yui, and Y. Yamachi, "Green Concrete Development: Evidence from Waste Concrete and Hemicellulose Utilization," *Open Journal of Civil Engineering*, vol. 14, pp. 587–601, 2024. https://doi.org/10.4236/ojce.2024.144032

[7] J. Kim, N. Nciri, A. Sicakova, and N. Kim, "Characteristics of waste concrete powders from multi-recycled coarse aggregate concrete and their effects as cement replacements," *Constr Build Mater*, vol. 398, p. 132525, Sep. 2023. https://doi.org/10.1016/j.conbuildmat.2023.132525

[8] L. Klaai, D. Hammiche, A. Boukerrou, and V. Pandit, "Thermal and structural analyses of extracted cellulose from olive husk," *Mater Today Proc*, vol. 52, pp. 104–107, 2022. https://doi.org/10.1016/j.matpr.2021.10.498

[9] W. Al Sekhaneh, A. Shiyyab, M. Arinat, and N. Gharaibeh, "Use of ftir and thermogravimetric analysis of ancient mortar from the church of the cross in gerasa (Jordan) for conservation purposes," *Mediterranean*

Archaeology and Archaeometry, vol. 20, no. 3, pp. 159–174, 2020. https://doi.org/10.5281/zenodo.4016073

[10]   R. Ylmén, U. Jäglid, B. M. Steenari, and I. Panas, "Early hydration and setting of Portland cement monitored by IR, SEM and Vicat techniques," *Cem Concr Res*, vol. 39, no. 5, pp. 433–439, May 2009. https://doi.org/10.1016/J.CEMCONRES.2009.01.017

[11]   V. Emmanuel, B. Odile, and R. Céline, "FTIR spectroscopy of woods: A new approach to study the weathering of the carving face of a sculpture," *Spectrochim Acta A Mol Biomol Spectrosc*, vol. 136, pp. 1255–1259, Feb. 2015. https://doi.org/10.1016/j.saa.2014.10.011

[12] R. Vârban *et al.*, "Comparative FT-IR Prospecting for Cellulose in Stems of Some Fiber Plants: Flax, Velvet

Leaf, Hemp and Jute," *Applied Sciences*, vol. 11, no. 18, p. 8570, Sep. 2021. https://doi.org/10.3390/app11188570

[13] I. Garcia-Lodeiro, G. Goracci, J. S. Dolado, and M. T. Blanco-Varela, "Mineralogical and microstructural alterations in a portland cement paste after an accelerated decalcification process," *Cem Concr Res*, vol. 140, p. 106312, Feb. 2021. https://doi.org/10.1016/j.cemconres.2020.106312

[14] V. H. J. M. dos Santos *et al.*, "Application of Fourier Transform infrared spectroscopy (FTIR) coupled with multivariate regression for calcium carbonate ($CaCO_3$) quantification in cement," *Constr Build Mater*, vol. 313, p. 125413, Dec. 2021. https://doi.org/10.1016/J.CONBUILDMAT.2021.125413

[15] M. Wu, B. Cui, H. Liu, and Z. Wang, "A lignin/castor oil-based polyamide autonomous self-healing composite materials," *Int J Biol Macromol*, vol. 305, p. 141159, May 2025. https://doi.org/10.1016/j.ijbiomac.2025.141159

[16] T.-P. Wang *et al.*, "Structures and pyrolytic characteristics of organosolv lignins from typical softwood, hardwood and herbaceous biomass," *Ind Crops Prod*, vol. 171, p. 113912, Nov. 2021. https://doi.org/10.1016/j.indcrop.2021.113912

[17] T. Rattana-amron, N. Laosiripojana, and W. Kangwansupamonkon, "Thermal oxidative degradation behavior

of extracted lignins from agricultural wastes: Kinetic and thermodynamic analysis," *Ind Crops Prod*, vol. 219, p. 119096, Nov. 2024. https://doi.org/10.1016/j.indcrop.2024.119096

[18] W. Fang, M. Alekhina, O. Ershova, S. Heikkinen, and H. Sixta, "Purification and characterization of kraft lignin," vol. 69, no. 8, pp. 943–950, 2015. https://doi.org/doi:10.1515/hf-2014-0200

[19] A. Bahari, A. Sadeghi-Nik, M. Roodbari, A. Sadeghi-Nik, and E. Mirshafiei, "Experimental and theoretical studies of ordinary Portland cement composites contains nano LSCO perovskite with Fokker-Planck and chemical reaction equations," *Constr Build Mater*, vol. 163, pp. 247–255, Feb. 2018. https://doi.org/10.1016/J.CONBUILDMAT.2017.12.073

[20] H. Ji *et al.*, "Facile synthesis, release mechanism, and life cycle assessment of amine-modified lignin for bifunctional slow-release fertilizer," *Int J Biol Macromol*, vol. 278, p. 134618, Oct. 2024. https://doi.org/10.1016/j.ijbiomac.2024.134618

[21] S. Yu, B. Qiu, Y. Jin, Y. Zhao, W. Luo, and X. Qi, "Efficient removal of lignin in tobacco stems with choline chloride-based deep eutectic solvents," *Ind Crops Prod*, vol. 226, p. 120634, Apr. 2025. https://doi.org/10.1016/j.indcrop.2025.120634

[22] Y. Zhu, H. Li, Q.-S. Zhao, and B. Zhao, "Effect of DES lignin incorporation on physicochemical, antioxidant and antimicrobial properties of carboxymethyl cellulose-based films," *Int J Biol Macromol*, vol. 263, p. 130294, Apr. 2024. https://doi.org/10.1016/j.ijbiomac.2024.130294

[23] W. Azuma, S. Nakashima, E. Yamakita, H. R. Ishii, and K. Kuroda, "Water retained in tall Cryptomeria japonica leaves as studied by infrared micro-spectroscopy," *Tree Physiol*, vol. 37, no. 10, pp. 1367–1378, Oct. 2017. https://doi.org/10.1093/treephys/tpx085

[24] Q. Xia, H. Peng, L. Yuan, L. Hu, Y. Zhang, and R. Ruan, "Anionic structural effect on the dissolution of arabinoxylan-rich hemicellulose in 1-butyl-3-methylimidazolium carboxylate-based ionic liquids," *RSC Adv*, vol. 10, no. 20, pp. 11643–11651, 2020. https://doi.org/10.1039/C9RA10108J

[25] J. Kim, N. Nciri, A. Sicakova, and N. Kim, "Characteristics of waste concrete powders from multi-recycled coarse aggregate concrete and their effects as cement replacements," *Constr Build Mater*, vol. 398, p. 132525, Sep. 2023. https://doi.org/10.1016/j.conbuildmat.2023.132525

[26] T. M. N. Tran, P. M.N., D.-W. Lee, and J. Song, "Enhanced mechanical and thermal properties of green PP composites reinforced with bio-hybrid fibers and agro-waste fillers," *Adv Compos Hybrid Mater*, vol. 8, no. 2, p. 224, Apr. 2025. https://doi.org/10.1007/s42114-025-01277-2

[27] L. M. Flórez-Pardo, A. González-Córdoba, and J. E. López-Galan, "Evaluation of different methods for efficient extraction of hemicelluloses leaves and tops of sugarcane," *Dyna (Medellin)*, vol. 85, no. 204, pp. 18–27, Jan. 2018. https://doi.org/10.15446/dyna.v85n204.66626

[28] M. Szymańska-Chargot *et al.*, "The Influence of High-Intensity Ultrasonication on Properties of Cellulose Produced from the Hop Stems, the Byproduct of the Hop Cones Production," *Molecules*, vol. 27, no. 9, p. 2624, Apr. 2022. https://doi.org/10.3390/molecules27092624

[29] Q. Peng, G. Ormondroyd, M. Spear, and W.-S. Chang, "The effect of the changes in chemical composition due to thermal treatment on the mechanical properties of Pinus densiflora," *Constr Build Mater*, vol. 358, p. 129303, Dec. 2022. https://doi.org/10.1016/j.conbuildmat.2022.129303.

[30] J. Zhuang, M. Li, Y. Pu, A. Ragauskas, and C. Yoo, "Observation of Potential Contaminants in Processed Biomass Using Fourier Transform Infrared Spectroscopy," *Applied Sciences*, vol. 10, no. 12, p. 4345, Jun. 2020. https://doi.org/10.3390/app10124345

Emerging Research in Materials for Environment, and Civil Infrastructure - GeoME 5.5   Materials Research Forum LLC
Materials Research Proceedings 58 (2026) 156-163   https://doi.org/10.21741/9781644903933-21

Integrating phase change materials for eco-friendly construction: Optimal positioning in building envelopes under Drâa-Tafilalet (Morocco) climate conditions

Azzeddine ELGHOMARI[1,a *]and Amine TILIOUA[1,b]

[1]Research Team in Thermal and Applied Thermodynamics (2.T.A.), Faculty of Science and Technology, Moulay Ismaïl UniversityErrachidia, 52000, Morocco.

[a]azzeddine.elghomari@gmail.com, [b]a.tilioua@umi.ac.ma

Keywords: Phase Change Materials (PCM), Energy Efficiency, Thermal Insulation, Composite Walls, Cement-Polystyrene, Thermal Comfort

Abstract. This study uses a laboratory-scale thermal enclosure with modular sidewalls to examine how phase change materials (PCMs) affect the thermal behavior of cement–polystyrene composite walls. Optimizing PCM placement within the wall structure for better heat control and energy efficiency is the primary goal. Detailed analyses of surface temperature evolution and heat-flux behavior confirm the results, which show that placing the PCM layer next to the heat source greatly increases latent heat utilization, improving thermal performance. Among the variables taken into account were phase change temperature, thermal load intensity, and heat transfer rate. Adding PCMs to the wall assembly considerably reduces surface temperature and heat transfer, according to experiments conducted at 41 °C, which mimics summer temperatures in Errachidia in the Drâa-Tafilalet region of southeast Morocco. These results demonstrate the effectiveness of PCM-based insulation systems in reducing energy demand in building envelopes and improving thermal efficiency.

Introduction

Because of the country's rapid urbanization and population growth, the building industry in Morocco has become the most energy-intensive of all industries. Buildings in Morocco use 25% of the energy and 30% of the CO_2 emissions that come from energy use [1]. Morocco aims to make buildings 30% more energy efficient over the next ten years, with 2015 as the baseline year. The main objective is to improve the envelope of the building to help decrease its cooling load. Conventional construction materials, particularly cement-based mortars, are not only low in thermal insulation capacity but also heavily involved in the greenhouse gas emissions cycle during both production and operational phases, thereby exacerbating the environmental problems associated with the building industry. In this light, there has been a shift toward the use of environmentally sustainable alternative materials with good thermal insulating properties. The natural materials have been one of the main options considered in this aspect as they are such a renewable resource with nature's recycling, low price, and favorable thermal behavior. Among the innovative answers being investigated, PCMs have been identified as exceptionally promising ones. During the phase transition, particularly between solid and liquid phases, PCMs can store and release thermal energy. Often this energy storage mechanism positively boosts the indoor thermal comfort by lessening dependence on active climate control systems such as heating and air conditioning, regulating humidity, and reducing temperature fluctuations [2, 3]. Environmental conditions, inherent material qualities, and most importantly the way in which PCMs are integrated into building components all have an impact on how successful they are [4,5]. Although numerous studies have recommended embedding PCMs into walls, floors, or ceilings, a consensus on the ideal placement to enhance thermal performance has yet to be established [6–8]. Numerous

Emerging Research in Materials for Environment, and Civil Infrastructure - GeoME 5.5   Materials Research Forum LLC
Materials Research Proceedings 58 (2026) 156-163   https://doi.org/10.21741/9781644903933-21

configurations, such as the usage of PCMs in exposed brick, panels, or plasterboards, have been the subject of experimental and numerical evaluations [9–14]. Stressing how important dynamic thermal reactions and thermophysical factors are. Recent research indicates that phase change materials (PCMs) installed on the interior side of walls are more effective at reducing peak indoor temperatures than those on the exterior [15, 16]. Lagou et al. [17], Hu and Yu [18], and Tunçbilek et al. [19] say that PCMs on the inside of walls worked better in hotter places, while PCMs on the outside of walls usually saved less energy because they had to cool down more often.

This research explores the incorporation of PCMs into cement polystyrene composite walls to achieve improved thermal management and optimized energy performance. The primary goal is to demonstrate the influence of PCMs placed at various portions of the wall using a laboratory-scale thermal enclosure. In this study, the location that can extract latent heat to reduce both surface temperature and heat transmission while increasing total thermal inertia will be found. This supports the creation of durable passive buildings that keep indoor conditions comfortable while reducing energy use for climate control in hot regions.

Materials and methods
Materials used. Phase Change Materials (PCMs) offer a more sustainable method of improving the thermal behavior of buildings than traditional building materials, which store energy based solely on changes in sensible heat. With PCMs, large amounts of energy are collected and released under relatively constant conditions by phase shifting from solid to liquid. PCMs can store up to 18 times as much energy per unit volume as traditional materials when they are the proper size. PCMs can thereby improve passive cooling, lower the quantity of active (mechanical) heating and cooling, and effectively control indoor temperatures. PCMs are easily integrated into any alternative building design that places a high priority on sustainability and energy efficiency because of their non-linear thermal behavior, which results from the accumulation of thermal energy using latent and sensible heat.

$$\Delta Q = C \times \Delta T = m \times c \times \Delta T \tag{1}$$

A material's thermal capacity C is determined by the product of its mass m and specific heat c, with heat storage taking place over a temperature range ΔT. Sensible heat is a common method of energy storage that is dependent on temperature fluctuations as well as the characteristics of the material. In contrast, PCMs store energy through latent and perceptible heat. As shown in (Fig. 1), PCMs take up or release substantial amounts of heat at an almost constant transition temperature T_m, as they shift between their solid and liquid states. A temperature plateau that occurs from this suggests latent heat storage. Furthermore, PCMs show little change in volume throughout the transition, which increases their dependability in thermal applications.

$$\Delta Q = \Delta H \tag{2}$$

The mechanism of buffering permits PCMs to manage temperature by absorbing heat during periods of extreme heat outdoors and in turn giving it off during cooling weather. In the above phase-change temperature range, PCMs behave as if they are normal substances and thus would draw sensible heat. The dual storage of heat leads to both temperature control and energy efficiency. The energy corresponding to the phase change is determined by the enthalpy change ΔH. A PCM stores a considerable amount of thermal energy as it undergoes a transition between solid and liquid states. Such materials are commonly classified within the broader category of thermal energy storage systems. PCMs transform into an endothermic phase, in which they absorb a great amount of heat without becoming any hotter and revert to solid before turning to liquid, under heat exceeding their melting point. In an exothermic transformation, PCMs revert to solid upon reaching a temperature below the melting point, releasing the accumulated energy during melting. Due to the capability to cycle on and off states, PCMs have a high utility toward building up energy stores, which explains their importance in buildings that require to maintain a stable temperature. PCM technology has progressed significantly in the past four decades, with numerous research works aiming at building them to perfection in composition and heat properties. Kerosene-based waxes, and specially created eutectic mixes and also the kind known as salt hydrates have been extensively investigated; each exhibits unique merits based on how each is to be deployed.

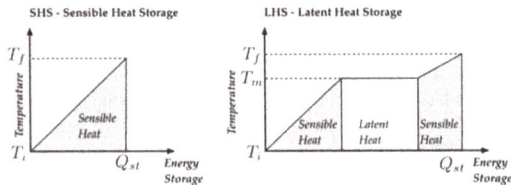

Fig. 1: Heat Absorption Behavior of Phase Change Materials under Thermal Charging.

This study presents a thermal insulation composite mortar made of cement and expanded polystyrene. It has low thermal conductivity and is lighter, making it a great choice for building projects that need to be energy-efficient. The improved synergy of the parts makes the thermal mass and resistance better, making it suitable for energy-efficient wall systems. The study also looks at how PCMs can be put together, focusing on how their arrangement in the wall construction affects the results. A control wall without PCM is used as a standard (Figs. 4). Thermal performance is measured by things like time lag, temperature amplitude damping, and surface heat flux. These things give us an idea of how much energy can be stored and how much thermal inertia there is.

In this test, 20% of the mortar was built using recycled polystyrene with particles in lengths between 0.5 and 1.0 cm (Fig. 3.c). The Microencapsulated MCP was produced as a soft panel which involved the use of a core matrix consisting of 40% polyethylene (a solid copolymer) and 60% kerosene, that was enveloped with two aluminum films of 75 μm thickness each (Figure 2). To guarantee the structural integrity and robustness of the panel, the perimeter was carefully sealed with aluminum tape having the same thickness as the panel. The final MCP composite obtained had a thickness of 5 mm and showed an in-plane density ranging from 250 to 1320 kg/m³.

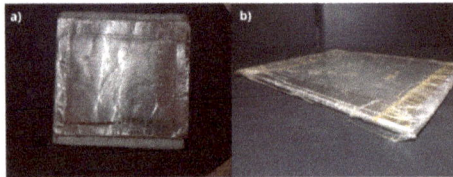

Fig 2: PCM panel with a thickness of 0.5 cm

Sample preparation. The mortar was modified through the inclusion of polystyrene waste, introduced specifically to enhance its thermal properties. The mortar was prepared using a cement-to-sand ratio of 1:3, in accordance with NM 10.1.046 standards. A single sample was prepared with a polystyrene volume fraction of 20%. To form a cohesive binder, water amounting to 50% of the cement mass was added to the cement sand composition. The polystyrene and mortar paste were subsequently mixed for a five-minute period to achieve a uniform and consistent mixture.

After that, the mixture was poured into molds that were $250 \times 250 \times 20$ mm³ in size. The sample was dried in the sun for a week, then wrapped in plastic to keep it dry (Fig. 3).

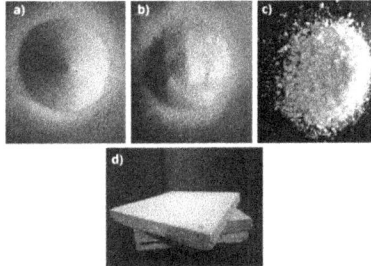

Fig. 3: Essential Components for EPS Lightweight Mortar Production.

Thermal conductivity. The experimental evaluation was conducted utilizing a bespoke, highly insulated thermal chamber engineered to replicate realistic heat transfer conditions with exceptional accuracy. The chamber has sidewalls that can be switched out, which lets you directly compare different wall compositions in the same environmental conditions. Thermal measurements were taken with an estimated uncertainty of less than 10% [20]. Fig. 4 shows the test configuration. The sidewalls each had a square 210×210 mm² cutout, which was thermally sealed during testing with a polystyrene panel measuring 5 cm in thickness to curb unintended heat dissipation. An internal heat source a100 W incandescent bulb encased in a black box was used to maintain an internal temperature near 41°C . The ambient environment was kept steady at approximately 21°C using an external cooling unit. Type K thermocouples were connected to a data logger to track the temperature on the inner and outer faces of the walls, with readings taken every two minutes. For each test, the wall surface was exposed to the heat source for 30 minutes at a fixed distance of 15 cm. As a result of this experimental setup, the placement and configuration of PCMs were evaluated, supporting the development of energy-efficient and environmentally friendly building envelope designs.

In the experimental configuration, each sample sized $250 \times 250 \times 25$ mm³ was placed on one of the sidewalls of the insulated test chamber, referred to as the thermal house. There were four thermocouples labelled T1 through T4 used to monitor temperature variations: one within the thermal chamber, one on each surface of the test sample, and one outside the enclosure (Fig. 5). Convective heat transfer is calculated using the following formula:

$$\varphi = h_i \times S(T_{i,wall} - T_{e,wall}) \tag{3}$$

Heat transfer through a material sample is fundamentally defined as the movement of thermal energy from a region of higher temperature to an adjacent region of lower temperature. This spontaneous flow of energy, governed by the Second Law of Thermodynamics, continues until thermal equilibrium is achieved across the specimen

$$\varphi = \lambda \times S \frac{(T_{i,wall} - T_{e,wall})}{e} \tag{4}$$

The calculation of thermal conductivity is carried out under the assumption that heat transfer occurs only in one dimension.

$$\lambda = h \times e \times \frac{(T_{i,air} - T_{i,wall})}{(T_{i,wall} - T_{e,wall})} \tag{5}$$

Emerging Research in Materials for Environment, and Civil Infrastructure - GeoME 5.5   Materials Research Forum LLC
Materials Research Proceedings 58 (2026) 156-163   https://doi.org/10.21741/9781644903933-21

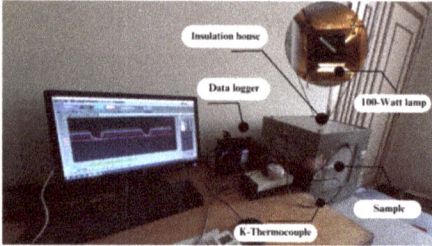

Fig. 4: Insulated thermal house.

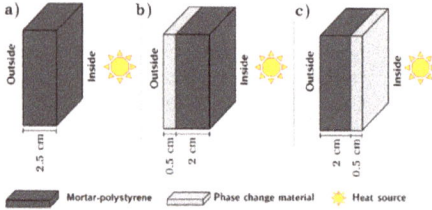

Fig. 5: Wall Configurations Schematic for Thermal Performance Analysis Using PCM Layer Positioning (a) No PCM (Control), (b) Interior PCM , and (c) External PCM.

Results

Following the manufacturer's guidelines for evaluating natural convection within confined spaces, the convective heat transfer coefficient (h_c) was set to 8.1 W/m²·K, as recommended in the technical documentation [21]. We then went ahead and measured the samples' thermal conductivity using the high-insulation home approach. A wall system's thermal behavior was examined in two primary scenarios: one in which Phase Change Material (PCM) was absent and the other in which PCM was positioned within or outside the thermal cavity in various configurations. The temperature changes over time at several monitored points, such as the ambient outdoor temperature and the inside air temperature, are depicted in (Fig. 6). The temperature profiles of the sample's external surface without PCM, with PCM inserted on the internal side, and with PCM embedded on the exterior side are shown in (Fig. 7).

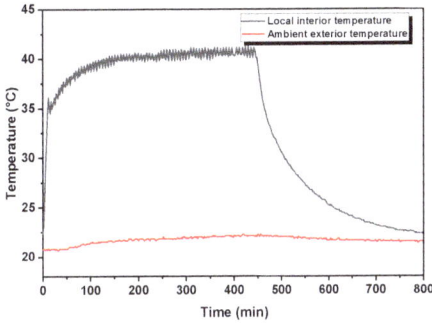

Fig. 6: Local interior and ambient exterior air temperatures of the control cavity.

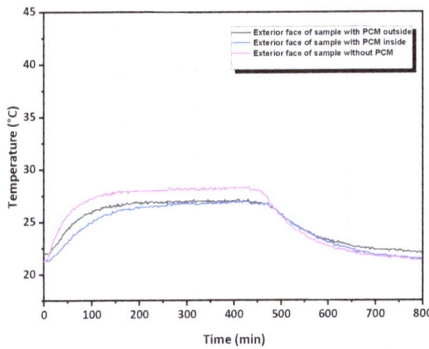

Fig. 7: External face temperature distribution of sample.

Discussion

The graph shows two different time periods: a charging period (0–470 minutes) when thermal energy is added, and a discharging period (470–800 minutes) when the system releases stored energy. The local interior temperature rises a lot during the charging phase, reaching a peak of about 41°C (Fig. 6). This shows that there are strong thermal gains. On the other hand, the samples with PCM show less extreme temperature increases. The sample with PCM inside the cavity has slightly lower temperature rises than the one with PCM outside, which suggests that PCM works better as a thermal buffer when it is inside. The sample without PCM has the highest temperature of all the wall surface measurements. This shows that there is no thermal storage moderation.

During the discharge phase, the samples containing PCM demonstrate a more gradual decrease in temperature, indicative of the latent heat release characteristics of PCM, which contributes to prolonged thermal comfort. Additionally, the configuration with PCM situated internally exhibits a more consistent temperature gradient compared to external PCM placement, further substantiating the notion that internal PCM positioning enhances the wall's thermal inertia more effectively. Throughout both phases, the ambient exterior temperature remains relatively stable at approximately 21–22°C, serving as a reliable benchmark. These findings confirm that PCM integration, particularly when applied to the internal side of the cavity, provides superior thermal regulation by mitigating peak interior temperatures and prolonging thermal comfort during the discharge phase.

Emerging Research in Materials for Environment, and Civil Infrastructure - GeoME 5.5   Materials Research Forum LLC
Materials Research Proceedings 58 (2026) 156-163   https://doi.org/10.21741/9781644903933-21

Conclusion

This study investigated two configurations: one incorporating PCMs and the other excluding them. The goal was to test how well cement polystyrene composite walls keep heat in by looking at how well they insulate. A custom-built thermal test chamber with collapsible walls was made to mimic heat gain and loss in a fictitious structure under laboratory conditions. The setup at high temperature enabled the artificial source to heat up and during lower temperature, heat to be slowly let out. One of the main goals was to find the best place for the PCM layer in the wall for the best thermal management by measuring changes in surface temperature. The experiment outcomes pointed out that placing the PCM layer nearer to the heat source improves thermal management since it takes in and gives out heat faster. Heat source temperature, outer wall surface, phase change temperature of the PCM, heat flow strength, and total heat storage capacity were some of the factors that affected the efficiency of the PCM. In the experiment, the heat source was set at 41°C to simulate average summer weather in the Drâa-Tafilalet area of Morocco. Under these conditions, walls with PCM showed a significant reduction in heat flow and surface temperature. The implications of this research are that walls with PCM can night-day cycle the indoor temperature and thus save energy due to thermal efficiency.

References

[1] ADEREE, Règlement Thermique de Construction au Maroc - Version simplifiée, 2014.

[2] Y. Zhang, G. Zhou, K. Lin, Q. Zhang, et H. Di. Application of latent heat thermal energy storage in buildings: State-of-the-art and outlook. Building and Environment, 42(6), (2007), 2197-2209. https://doi.org/10.1016/j.buildenv.2006.07.023

[3] H. Nasser, Thesis Innovative heat transfer enhancements for thermal energy storage systems based on phase change materials, 2019. https://etheses.whiterose.ac.uk/25192/

[4] Z. Ben Zaid, A. Tilioua, I. Lamaamar, O. Ansari, H. Souli, M. A. Hamdi Alaoui. An experimental study of the efficacy of integrating a phase change material into a clay-straw wall in the Draa-Tafilalet Region (Errachidia Province), Morocco. Journal of Building Engineering, 32, (2020), 101670.https://doi.org/10.1016/j.jobe.2020.101670.

[5] U. Berardi, A. A. Gallardo. Properties of concretes enhanced with phase change materials for building applications. Energy and Buildings, 199, (2019), 402-414. https://doi.org/10.1016/j.enbuild.2019.07.014.

[6] F. Souayfane, F. Fardoun, P.H. Biwole, Phase change materials (PCM) for cooling applications in buildings: a review. Energy and Buildings, 129, (2016), 396-431.

[7] A. Hasan, K.A. Al-Sallal, H. Alnoman, Y. Rashid, S. Abdelbaqi, Effect of phase change materials (PCMs) integrated into a concrete block on heat gain prevention in a hot climate, Sustainability, 8(10), (2016), 1009. https://doi.org/10.3390/su8101009.

[8] A. Gounni, M. El Alami, The optimal allocation of the PCM within a composite wall for surface temperature and heat flux reduction: an experimental Approach. Applied Thermal Engineering, 127, (2017), 1488-1494. https://doi.org/10.1016/j. applthermaleng.2017.08.168.

[9] V. Sá, M. Azenha, H. De Sousa, A. Samagaio . Thermal enhancement of plastering mortars with Phase Change Materials: Experimental and numerical approach. Energy and Buildings. 49, (2012), 16-27. https://doi.org/10.1016/j.enbuild.2012.02.031

[10] P. Schossig, H.M. Henning, S. Gschwander, T. Haussmann, Micro-encapsulated phase-change materials integrated into construction materials. Solar Energy Materials and Solar Cells. 89(2-3), (2005), 297-306. https://doi.org/10.1016/j.solmat.2005.01.017.

[11] F. Ascione, N.Bianco, R. F.De Masi, F.De' Rossi, G. P.Vanoli. Energy refurbishment of existing buildings through the use of phase change materials: Energy savings and indoor comfort in the cooling season. Applied Energy, 113, (2014), 990-1007. https://doi.org/10.1016/j.apenergy.2013.08.045

[12] Q. Al-Yasiri, M. Szabó, Case study on the optimal thickness of phase change material incorporated composite roof under hot climate conditions. Case Studies in Construction Materials, 14, (2021), e00522. https://doi.org/10.1016/j.cscm.2021.e00522.

[13] R. Vicente, T. Silva, Brick masonry walls with PCM macrocapsules: An experimental approach. Applied Thermal Engineering, 67(1-2), (2014), 24-34. http://dx.doi.org/10.1016/j.applthermaleng.2014.02.069

[14] Castell, I. Martorell, M. Medrano, G. Pérez, L.F. Cabeza, Experimental study of using PCM in brick constructive solutions for passive cooling. Energy and Buildings, 42(4), (2010), 534-540. https://doi.org/10.1016/j.enbuild.2009.10.022

[15] D. Mazzeo, G. Oliveti, A. de Gracia, J. Coma, A. Solé, L.F. Cabeza, Experimental validation of the exact analytical solution to the steady periodic heat transfer problem in a PCM layer, Energy. 140, (2017), 1131-1147. https://doi.org/10.1016/ j.energy.2017.08.045.

[16] D. Mazzeo, G. Oliveti, Thermal field and heat storage in a cyclic phase change process caused by several moving melting and solidification interfaces in the layer. International Journal of Thermal Sciences, 129, (2018), 462-488. https://doi.org/10.1016/j.ijthermalsci.2017.12.026.

[17] Lagou, A. Kylili, J. Sadauskiene, P.A. Fokaides. Numerical investigation of phase change materials (PCM) optimal melting properties and position in building elements under diverse conditions. Construction and Building Materials. 225, (2019), 452-464. https://doi.org/10.1016/j.conbuildmat.2019.07.199.

[18] J. Hu, X. Yu. Adaptive building roof by coupling thermochromic material and phase change material: Energy performance under different climate conditions. Construction and Building Materials .262, (2020), 120481. https://doi.org/10.1016/j.conbuildmat.2020.120481

[19] E.Tunçbilek, M.Arıcı, M. Krajčík, S.Nižetić, H.Karabay. Thermal performance-based optimization of an office wall containing PCM under intermittent cooling operation. Applied Thermal Engineering. 179, (2020), 115750. https://doi.org/10.1016/j.applthermaleng.2020.115750

[20] D. Belatrache, S.Bentouba, N. Zioui, M. Bourouis. Energy efficiency and thermal comfort of buildings in arid climates employing insulating material produced from date palm waste matter. Energy, 283, (2023), 128453. https://doi.org/10.1016/j.energy.2023.128453.

[21] P2360300 PHYWE Series of Publications, Laboratory Experiments, Physics, PHYWE SYSTEME GMBH & Co.KG, Göttingen.

Emerging Research in Materials for Environment, and Civil Infrastructure - GeoME 5.5   Materials Research Forum LLC
Materials Research Proceedings 58 (2026) 164-171   https://doi.org/10.21741/9781644903933-22

Development of ceramic membranes based on Draa-Tafilalet clay for Malachite Green retention in water filtration

Mohammed Chrachmy[1,a] *, Ayoub Souileh[2,b], Rajae Ghibate[3,c], Achraf Mabrouk[4,d],
Mohamed Ech-Chykry[1,e], Anjoud Harmouzi[5,f], Najia El Hamzaoui[6,g],
Meryem Ben Baaziz[1,h], Hassan Ouallal[1,I], and Mohamed Azrour[1,j]

[1]Laboratory of Materials Engineering for the Environment and Natural Resources, Faculty of Sciences and Technologies, Moulay Ismail University of Meknes, Errachidia, Morocco

[2]L3GIE, Mohammadia Engineering School, Mohammed V University in Rabat, Morocco

[3]Laboratory of Physical Chemistry, Materials and Environment, Faculty of Sciences and Technologies, Moulay Ismail University of Meknes, Errachidia, Morocco

[4]Laboratory of Agri-food and Health (LAFH), Faculty of Sciences and Techniques, Hassan 1st University, Settat, Morocco

[5]Laboratory of Natural Resources and Sustainable Development, Faculty of Sciences, Ibn Tofail University, Kenitra, Morocco

[6]Higher Institute of Nursing and Health Professions of Fez-Meknes. Regional Directorate of Health Fez-Meknes, Meknes, Morocco

[a]chrachmy@gmail.com, [b]Ayoub.souileh@research.emi.ac.ma, [c]rajae.ghibate@gmail.com, [d]chraf.mab@gmail.com, [e]mohamed.echykry@gmail.com, [f]nojoud.harmouzi@acheivemency.com, [g]najiaelhamzaoui1@gmail.com, [h]benbaazizmeryem@yahoo.fr, [I]h.ouallal@umi.ac.ma, [j]mohamedazrour@yahoo.fr

Keywords: Clay, Ceramic Membrane, Draa-Tafilalet, Physicochemical Characterization, Water Filtration

Abstract. Water pollution by synthetic dyes such as Malachite Green (MG) poses a significant environmental challenge, particularly in developing countries. This study aims to create eco-friendly ceramic membranes from natural clay sourced from the Draa-Tafilalet region of Morocco to filtrate MG-contaminated water. The membranes were fabricated via a dry pressing method using varying proportions (0–12 wt%) of date palm pits as an organic pore-forming agent. The Physicochemical characterizations, revealed that increasing the additive content enhances membrane porosity while decreasing mechanical strength. With the optimal formulation identified at 6 wt% and sintering at 1000°C. The elaborated membrane was applied to the frontal microfiltration of MG dye solution. UV-Vis spectroscopic analysis confirmed high dye removal efficiency, with more than 70% retention after 200 minutes of filtration. Meanwhile, the permeate flow gradually decreased due to fouling, revealing a trade-off between filtration efficiency and flow rate. The results demonstrate that natural clay-based ceramic membranes, modified with agricultural waste, provide an environmentally friendly and cost-efficient approach for dye removal in water treatment systems.

Introduction

Water bodies in many developing regions continue to face contamination from industrial dyes, and Malachite Green (MG) is one of the compounds that often causes serious concern. This dye is widely used, mainly in textile processing and sometimes in aquaculture, and several studies have already shown its harmful effects on living organisms. Its persistence in water makes the situation more complicated, as MG does not degrade easily and can remain for long periods in natural

Emerging Research in Materials for Environment, and Civil Infrastructure - GeoME 5.5   Materials Research Forum LLC
Materials Research Proceedings 58 (2026) 164-171   https://doi.org/10.21741/9781644903933-22

environments [1]. Different treatment methods have been tested to remove MG. Some authors worked with solar irradiation [2] or sonochemical degradation [3], whereas others tried biological or advanced oxidation processes [4,5]. Many of the current treatments for eliminating MG from water have already been investigated, and while some of them can be effective under particular circumstances, their actual application is usually still challenging. Many of times they can produce secondary residues needing extra handling, therefore they only function effectively over small pH or temperature ranges. Particularly in areas where treatment facilities are already constrained, the financial expense of these procedures is another constraint. For these reasons, several research groups have been looking at alternatives that would be less expensive in terms of resources and simpler to execute. For water treatment recently, membrane-based filtration has shown itself to be a really good option. Retaining various categories of pollutants from big suspended particles to dissolved organic molecules such dyes [7] , depends on the kind of membrane utilized. Because of their toughness, ceramic membranes stand out among the several membrane families. One benefit of their ability to be created from locally accessible clay materials is reduced production costs and increased use of local raw resources [8]. They can survive high temperatures, harsh chemical environments, and multiple cleaning procedures without sacrificing their structural stability. Sometimes, organic additives are added during preparation to adjust the pore structure and improve permeability [9]. Considering these aspects, ceramic membranes made from local clay could represent a promising option for treating dye-contaminated wastewater. The aim of this study is to prepare and test ceramic membranes produced from clay collected in the Draa-Tafilalet region of Morocco. The work attempts to provide a treatment approach that is efficient but also accessible for areas where technological resources remain limited.

Material and methods
Raw material. The clay material employed in this investigation was collected from Draa-Tafilalet region of Morocco particularly in the Es-Sifa rural area. As shown in Fig. 1. The geographic coordinates of the sampling zone range from 4°18'66"W to 4°20'00"W in longitude and from 31°19'14"N to 31°20'34"N in latitude [10].

To guarantee the quality and representativity of the sample while avoiding surface contamination, clay was taken from multiple points at depths exceeding 50 cm. To enhance the porosity of the membranes developed in this study, date palm pits, were used as an organic additive. These pits were previously washed, dried, ground, and sieved to retain only particles smaller than 315 μm.

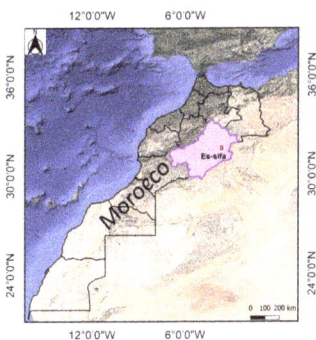

Fig. 1: Location map of the sampling site

Emerging Research in Materials for Environment, and Civil Infrastructure - GeoME 5.5   Materials Research Forum LLC
Materials Research Proceedings 58 (2026) 164-171   https://doi.org/10.21741/9781644903933-22

Ceramic Membranes Elaboration Process. Fig. 2 illustrates the different steps in the elaboration of membranes. The dry uniaxial pressing was employed to fabricate the ceramic membranes, with varying the percentages of date palm pits.

Discshaped pellets, measuring approximately 2 mm in thickness and 40 mm in diameter, were formed by uniaxial dry pressing using a hydraulic press under a pressure ranging from 10 to 15 tons. Pits powder was incorporated into the clay mixtures at proportions from 0 to 12 wt%.

This pits powder is eliminated in the thermal treatment. Contributing to the porosity within the ceramic matrix, following shaping, the pellets non-sintered were subjected to a controlled thermal treatment consisting of successive heating steps at 4°/min to 100, 250, 500, and finally 1000 °C, at each temperature a 2 hour holding period was applied.

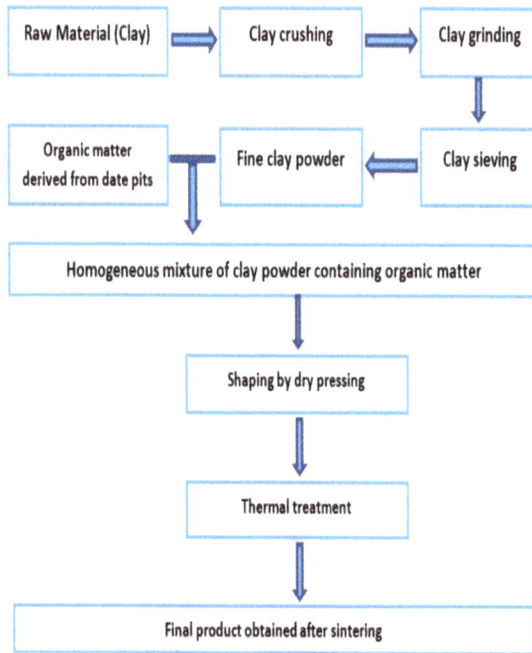

Fig. 2: Elaboration Process of Ceramic Membranes.

Characterization of Membranes. The characterization of the membranes was carried out by assessing their porosity and flexural strength.

The porosity was evaluated through the water absorption technique following the ISO 10545-3 standard. In this method, the dry mass (M_{sec}), the saturated mass after immersion in water (M_{sat}) and the hydrostatic mass of the sample (M_h) were measured. The porosity P (%) was then calculated using Eq. (1) [12].

$$P = \frac{M_{sat} - M_{sec}}{M_{sat} - M_h} * 100 \qquad (1)$$

The mechanical strength was evaluated by a three-point bending test. The specimens with dimensions of 1 cm in thickness, 3 cm in width, and 7 cm in length were fabricated under the same shaping and sintering conditions as the final membranes.

During the test, a gradually increasing load (F) was applied at the midpoint of the specimen, supported at both ends, until fracture occurred. The distance between the two supports was denoted by d, the sample thickness bye, and the sample width by l. The flexural strength σ (Mpa) was calculated according to Eq. (2) [11]:

$$\sigma = \frac{3 \times F \times d}{2 \times l \times e^2} \qquad (2)$$

Evaluation of Permeate flux. The permeate flux was conducted using a laboratory-scale filtration process as illustrated in Fig. 3.

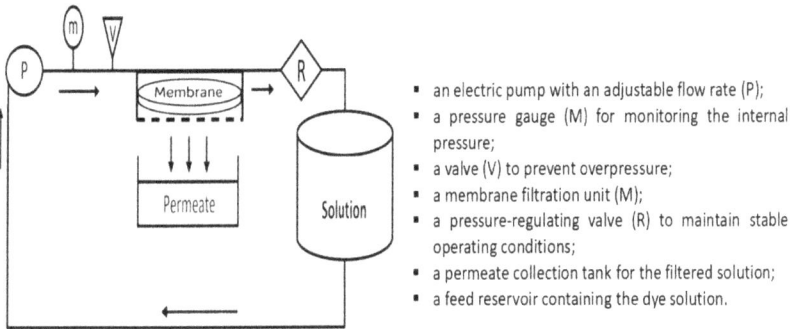

- an electric pump with an adjustable flow rate (P);
- a pressure gauge (M) for monitoring the internal pressure;
- a valve (V) to prevent overpressure;
- a membrane filtration unit (M);
- a pressure-regulating valve (R) to maintain stable operating conditions;
- a permeate collection tank for the filtered solution;
- a feed reservoir containing the dye solution.

Fig. 3: Schematic of the micro-pilot for frontal filtration

The permeate flux (D) of the ceramic membranes was evaluated from the volume of filtrate collected during the filtration experiment and calculated according to Eq. (3), where Vis the volume of permeate collected (L), t represents the duration of filtration (h), S corresponds to the effective membrane surface area (m²); and D expresses the permeate flux in liters per hour per square meter of membrane surface ($L \cdot h^{-1} \cdot m^{-2}$).

$$D = \frac{V}{t \times S} \; (L/h.m^2) \qquad (3)$$

Results and Discussion

Clay particles characterization. In our previous works [13–15], the clay under study was described.

X-ray diffraction studies showed quartz to be the most abundant mineral phase together with lesser phases like calcite, dolomite, and kaolinite.

The variety of this minerals suggests a sophisticated crystalline arrangement that supports the emergence of porosity after heat treatment.

Heating showed a few changes that thermal analysis discovered. The removal of physically adsorbed water accounted for the first mass loss seen at low temperatures. Organic matter was seen to break down at moderate temperatures, next to a low-intensity endothermic signal pointing

Emerging Research in Materials for Environment, and Civil Infrastructure - GeoME 5.5   Materials Research Forum LLC
Materials Research Proceedings 58 (2026) 164-171   https://doi.org/10.21741/9781644903933-22

to a low kaolinite concentration. The α - β phase transition of quartz happened at higher temperatures together with the decarbonation of carbonate phases (calcite and dolomite). At last, a structural change matching the transformation of metakaolinite into spinel-type alumino-silicate phases and amorphous silica was observed.

FTIR Spectroscopy supported the presence of functional groups usually found in natural clays. Hydroxyl (–OH) vibrations related with Al, Mg, or Fe were given a strong absorption band. A wide band suggested adsorbed water; clear absorption bands confirmed calcite and quartz. Deformation vibrations associated with silicate (Si-O) bonds were also seen.

Because it can create a porous structure upon sintering, this clay is a strong candidate for membrane preparation depending on its physiochemical and mineralogical characteristics.

Optimization of the Membrane Formulation.

A major characteristic that has to be closely taken into account when creating ceramic membranes is porosity since it greatly affects both their mechanical behavior and permeability.

To boost the porosity of the produced membranes, date palm pits were combined with the clay at weight ratios between 0 and 12%. Fig. 4 shows how additive level affects porosity and mechanical strength over time. The results clearly show a strong inverse correlation between these two qualities: the porosity of the membranes increases as the proportion of date palm pits rises, therefore reducing their mechanical strength. The creation of pores during the thermal decomposition of organic material causes this behavior. The burning of date palm pits releases volatile chemicals during sintering, which expulsion generates voids in the ceramic matrix. Consequently, the pore volume seen is precisely related to the amount of organic material added. Furthermore aiding to raise the energy efficiency of the sintering process is this exothermic combustion process. Among all the examined compositions, the one containing 6% date palm pits and sintered at 1000 °C emerged as the best balance since it provided enough porosity without compromising the mechanical integrity of the membrane.

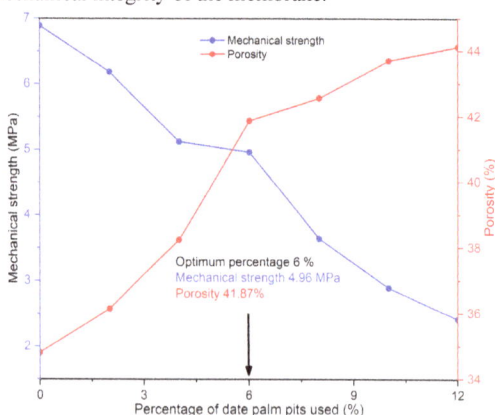

Fig. 4: Optimization of the organic additive

Application in Dye Solution Treatment

Properties of the Studied Dye. The absorption spectrum of the Malachite Green solution—a cationic dye from the triphenylmethane family—with a maximum absorption at 620 nm is presented in Fig. 5. Table 1 summarizes its important physico-chemical characteristics. Before and

Emerging Research in Materials for Environment, and Civil Infrastructure - GeoME 5.5   Materials Research Forum LLC
Materials Research Proceedings 58 (2026) 164-171   https://doi.org/10.21741/9781644903933-22

after the filtering process, this wavelength helped to establish the dye concentration in the solutions.

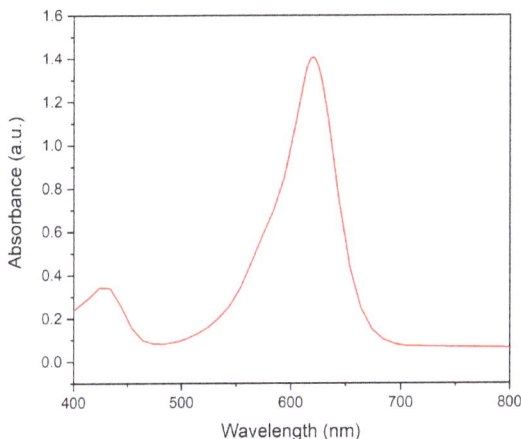

Fig. 5: UV-visible spectrum of Malachite Green

Table 1: MG physico-chemical properties

Name	Malachite green oxalate
Molecular Formula	C52H54N4O12
Physical state and color of the compound	Green crystalline powder
Molecular. Weight (g/mol)	927.02
Chemical structure	

Permeate Flow and Filtration Efficiency. Over time at a consistent pressure of 0.5 bar, Fig. 6 displays the filtration efficiency and permeate flow rate.

Initially with a flow rate of around 100 L/m²/h, the flow lowers over time as a result of probable membrane fouling or progressive blockage of the pores by retained particles. On the other hand, the filtration efficacy starts at about 10% and rises to over 70% by the finish of the 200-minute filtration process.

Emerging Research in Materials for Environment, and Civil Infrastructure - GeoME 5.5   Materials Research Forum LLC
Materials Research Proceedings 58 (2026) 164-171   https://doi.org/10.21741/9781644903933-22

The inverse link between flow and efficiency suggests that, as the membrane's permeability falls, its retention capacity increases probably because of the development of a cake layer on the surface that improves rejection of pollutants. These findings show how selectivity and permeability interact during extended filtration.

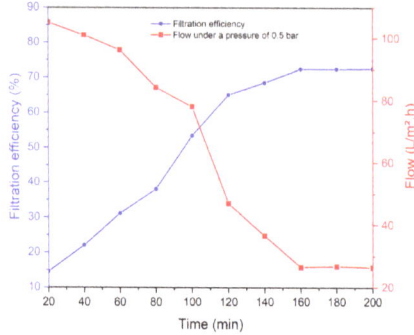

Fig. 6: Variation of filtration efficiency and permeate flow over time

Conclusion

Using date pits as a sustainable and easily accessible porogen, this study shows that ceramic membranes made from natural Moroccan clay can be effectively manufactured. The findings indicate an inverse association between porosity and mechanical strength; the composition comprising 6% by weight of kernel powder provided the ideal compromise between permeability and structural strength. After 200 minutes of operation at a constant pressure of 0.5 bar, the optimized membrane was microfiltrated with a malachite green solution, removing more than 70% of the dye. Although the permeate flow slowly dropped as a result of fouling events, dye retention efficiency improved across the filtration period, showing the membrane system's enhanced selectivity. All things considered, these findings show the promise of clay-based ceramic membranes infused with locally acquired organic compounds as a cheap and ecologically friendly technique for treating dye-contaminated wastewater.

References

[1] S. Srivastava, R. Sinha, D. Roy, Toxicological effects of malachite green, Aquat. Toxicol. 66 (2004) 319–329. https://doi.org/10.1016/j.aquatox.2003.09.008

[2] N. Nayak, S. Singha, J.P. Maity, P.P. Rath, T. Sahoo, T.R. Sahoo, Photocatalytic degradation of malachite green dye under solar light irradiation using ZnO and ZnO–TiO2 nanoparticles, J. Mater. Sci.: Mater. Electron. 35 (2024) 310. https://doi.org/10.1007/s10854-024-12066-w.

[3] O. Moumeni, O. Hamdaoui, C. Pétrier, Sonochemical degradation of malachite green in water, Chem. Eng. Process.: Process Intensif. 62 (2012) 47–53. https://doi.org/10.1016/j.cep.2012.09.011

[4] L.-N. Du, M. Zhao, G. Li, F.-C. Xu, W.-H. Chen, Y.-H. Zhao, Biodegradation of malachite green by Micrococcus sp. strain BD15: Biodegradation pathway and enzyme analysis, Int. Biodeter. Biodegr. 78 (2013) 108–116. https://doi.org/10.1016/j.ibiod.2012.12.011

[5] G.-Y. Lv, J.-H. Cheng, X.-Y. Chen, Z.-F. Zhang, L.-F. Fan, Biological decolorization of malachite green by Deinococcus radiodurans R1, Bioresour. Technol. 144 (2013) 275–280. https://doi.org/10.1016/j.biortech.2013.07.003

[6] A. Kausar, M. Iqbal, A. Javed, K. Aftab, Z.-H. Nazli, H.N. Bhatti, S. Nouren, Dyes adsorption using clay and modified clay: A review, J. Mol. Liq. 256 (2018) 395–407. https://doi.org/10.1016/j.molliq.2018.02.034

[7] M.S. Kanth, S.L. Sandhya Rani, V.K. Raja, Advancing ceramic membrane technology in chemical industries applications by evaluating cost-effective materials, fabrication and surface modification methods, Hybrid Advances 8 (2025) 100380. https://doi.org/10.1016/j.hybadv.2025.100380

[8] M.B. Asif, Z. Zhang, Ceramic membrane technology for water and wastewater treatment: A critical review of performance, full-scale applications, membrane fouling and prospects, Chem. Eng. J. 418 (2021) 129481. https://doi.org/10.1016/j.cej.2021.129481

[9] S. Lagdali, M. El-Habacha, G. Mahmoudy, M. Benjelloun, S. Ssouni, Y. Miyah, S. Iaich, M. Zerbet, Development and characterization of an asymmetrical flat microfiltration membrane based on natural phengite clay: Application as a pretreatment for raw seawater reverse osmosis desalination, J. Water Process Eng. 67 (2024) 106253. https://doi.org/10.1016/j.jwpe.2024.106253

[10] M. Chrachmy, R. Ghibate, N. El Hamzaoui, A. Ansari, A. Souileh, A. Harmouzi, M. Ben Baaziz, H. Ouallal, M. Azrour, Exploring Es-sifa clay as a high-performance adsorbent: experimental and density functional theory investigations, Scientific African, 29, e02806. 2025, https://doi.org/10.1016/j.sciaf. 2025.e02806

[11] M. Sawadogo, L. Zerbo, M. Seynou, S. Brahima, R. Ouedraogo, Technological properties of raw clay based ceramic tiles: Influence of talc, Sci. Stud. Res.15 (2014) 231–238.

[12] I. Demir, Effect of organic residues addition on the technological properties of clay bricks, Waste Manag. 28 (2008) 622–627. https://doi.org/10.1016/j.wasman.2007.03.019

[13] M. Chrachmy, R. Ghibate, N. El Hamzaoui, M. Lechheb, H. Ouallal, M. Azrour, Application of raw Moroccan clay as a potential adsorbent for the removal of malachite green dye from an aqueous solution: Adsorption parameters evaluation and thermodynamic study, Mater. Res. Proc. 40 (2024) 248– 259. https://doi.org/10.21741/9781644903117-27

[14] M. Chrachmy, M. Lechheb, H. Ouallal, N. El Hamzaoui, A. Souileh, M.Azdouz, M. Azrour, Improving thermal insulation and mechanical properties of building bricks made from Moroccan clay, Mater. Res. Proc. 40 (2024) 260–272. https://doi.org/10.21741/9781644903117-28

[15] H. Ouallal, M. Chrachmy, N. El Hamzaoui, M. Lechheb, R. Ghibate, H. Elmarjaoui, M. Azrour, Insight on the natural Moroccan clay valorization for malachite green adsorption: kinetic and isotherm studies, Mater. Res. Proc. 40 (2024) 273–283. https://doi.org/10.21741/9781644903117-2

Keyword Index

3D Modelling 116
3D Printed Concrete 62

Abaqus 47
Architecture 86
Artificial Intelligence 9, 24

BIM 108
Binder 32
Bio-Based Reinforcement 139
Bracing Systems 1
Buckling 17

Carbon Steel 17
Cellular Beams 17
Cellulose 147
Cement Replacement 78
Cement-Polystyrene 156
Ceramic Membrane 164

Clay 99, 164
Climatic Conditions 130
Combined NDT Methods 92
Composite Walls 156
Compressed Earth Block 40
Compressive Strength 9, 78
Concrete Strength Variability 92
Concrete 9
Conservation 116
Core Sampling Strategies 92
Cultural Heritage 108, 116

Date Palm Fibers 54
Digital Fabrication 62
Digital Heritage 123
Digital Survey 116, 123

Digital Technologies 108
Draa-Tafilalet 164

Earth Blocs 32
Earth 86
Earthen Architecture 123
Eco-Friendly Materials 62
Energy Efficiency 156
Energy Efficiency 99, 130
Ensemble Learning 9
Epoxy Matrix 70
Expansive Soils 139

FEM 17
Fibers 70
Fly Ash 24
FTIR 147

Geotechnical Issues 139
Gypsum Plaster 99
Gypsum 54

Haouz Valley 86
Hardness Test 70
Hemicellulose 147
Hydrothermal Behavior 32
In-Situ Concrete Strength 92

Lignin 147
Low-Carbon Concrete 62

Machine Learning Algorithms 9
Machine Learning 24
Masonry Structure 47
Mechanical Proprieties 70
Mechanical Resistance 32
Mechanical Strength 24
Mensiocronology 108

Morocco	86, 116, 123	Thermal Comfort	156
Mortar	78	Thermal Insulation	54, 156
		Thermal Performance	99, 130
Natural Fibers	139	Timber	86
Natural Pozzolan	78	Traditional Building Materials	130
Numerical Simulation	54		
		Ultrasonic Pulse Velocity (UPV)	92
Phase Change Materials (PCM)	156	Ultrasound	40
Photogrammetry	116, 123		
Physicochemical Characterization	164	Vernacular	86
Porosity	78	Vulnerability	40
Pozzolan	32		
Pozzolanic Materials	78	Water Filtration	164
Pushover Analysis	1		
Random Coring	92		
RC Structure	1		
Rebound Hammer Testing	92		
Recycled Concrete Powder	147		
Repeated Load Cycles	47		
Restoration	108		
Seismic Performance	86		
Seismic Vulnerability	1		
Seismic	40		
Simulation	70		
Soil Stabilization	139		
Solid Panel	47		
Stability Performance	47		
Stainless Steel	17		
Sugarcane Bagasse Ash	78		
Sustainable Alternatives	139		
Sustainable Concrete	24		
Sustainable Construction	62, 130		
Sustainable Materials	54		
Sustainable Materials	99		
Tensile Test	70		

About the Editors

Prof. BABA Khadija

KHADIJA BABA is a professor of civil engineering at the higher school of technology in Salé, Mohammed V University, Morocco, where she has been shaping young minds since 1995. As a PhD holder from Mohammadia school of engineers (EMI) in Rabat, she currently leads the soil mechanics, structures, and materials research team at the civil and environmental engineering laboratory and serves as a valued member of the UNESCO Chair on Education and Research in Urban and Bioclimatic Sustainable Architecture at the National School of Architecture in Rabat.

A recognized expert in sustainable construction materials, KHADIJA BABA has innovative research in industrial waste recovery and recycling for civil engineering applications. Her scholarly contributions include over 100 high-impact publications and several co-edited books. She serves on the editorial board of environmental sciences and sustainable development (ESSD) journal, where she contributes to advancing knowledge in sustainable engineering practices.

As the founder and chair of the international conference on geosciences and environmental management (GeoME), KHADIJA BABA has established a vital platform for global knowledge exchange. Her leadership extends to numerous research projects focusing on the valorisation of local construction materials, demonstrating her commitment to sustainable development and circular economy principles in civil engineering.

Prof. NOUNAH Abderrahman

ABDERRAHMAN NOUNAH holds a PhD in physical chemistry and materials science. he is a professor at Mohammed V University and director of the higher school of technology of Salé in Morocco. he is also the director of the laboratory of civil engineering and environment at Mohammadia school of engineering. his research focuses on the recovery and recycling of industrial waste in civil engineering, and he has published over sixty articles and book chapters. ABDERRAHMAN NOUNAH has participated in research projects on the valorization of local construction materials and serves as an expert evaluator at CNRST rabat. he is the vice president of ASMATEC, promoting education and research in materials science and construction technology.

Prof. OUADIF Latifa

Latifa OUADIF is a professor at the mineral engineering department of Mohammadia school of engineers in Rabat, Morocco. She is a member of the laboratory of applied geophysics, geotechnics, engineering geology, and environment (L3GIE) at the same institution. Her research focuses on rock mechanics, engineering geology, and environmental geology, among other areas. She has published over 100 articles and book chapters, co-edited multiple books, and actively teaches and supervises Ph.D. And undergraduate students. Latifa OUADIF is a co-founder and co-chair of the international conference on geosciences and environmental management (GeoME).

Prof. BAHI LAHCEN

LAHCEN BAHI is a distinguished academic and leading expert in applied geosciences and environmental engineering with over 38 years of experience at Mohammed V University's Mohammadia School of Engineers (EMI) in Rabat. Former director of both the doctoral studies center (CeDoc) and the civil engineering, water & geosciences research center at EMI, he currently serves as the founding director of the L3GIE laboratory, a position he has held since 1986. Trained at the university of Strasbourg (state doctorate in geophysics) and EMI (engineering diploma in geological engineering), prof. BAHI specializes in sustainable mining, applied geophysics, and environmental impact assessment, with particular focus on bridging academic research with industrial applications. His pioneering work has fostered strategic R&D partnerships with major national organizations. While his leadership has guided 50+ PhD candidates to completion. Recognized for developing methodologies that integrate sustainability into industrial projects, he actively contributes to scientific committees and policy development at national and international levels. Prof. BAHI continues to advance geo-engineering solutions that address Morocco's pressing environmental and resource management challenges while mentoring the next generation of scientists and engineers.

Prof. EL RHAFFARI Younes
Younes EL RHAFFARI holds a PhD in Physical Sciences, Mechanics, And Energy at the Faculty of Sciences, Mohamed V University, Rabat; He is a Professor at Mohammed V University of Rabat, Higher School of Technology of Salé; He is permanent member of the Laboratory of Civil Engineering and Environment at Mohammadia School of Engineering; His research activities are mainly focused on characterization of construction materials of historical monuments; he has published a several articles and book chapters in indexed journals; Younes EL RHAFFARI has participated in several projects and research conferences on construction materials; He is member of the board of the Miftah Essaad Foundation for the Intangible Capital of Morocco; He is member of the International Council on Monuments and Sites (ICOMOS); He is member of standing committee of the RIPAM (international meetings of Mediterranean architectural heritage).